Recommended Health-Based Limits in Occupational Exposure to Heavy Metals

Report of a
WHO Study Group

World Health Organization
Technical Report Series
647

World Health Organization, Geneva 1980

ISBN 92 4 120647 0

The designations employed and the presentation of the material in this publication do not imply the expression of any opinion whatsoever on the part of the Secretariat of the World Health Organization concerning the legal status of any country, territory, city or area or of its authorities, or concerning the delimitation of its frontiers or boundaries.

The mention of specific companies or of certain manufacturers' products does not imply that they are endorsed or recommended by the World Health Organization in preference to others of a similar nature that are not mentioned. Errors and omissions excepted, the names of proprietary products are distinguished by initial capital letters.

PRINTED IN SWITZERLAND

80/4677 — La Concorde — 8000

CONTENTS

WHO STUDY GROUP ON RECOMMENDED HEALTH-BASED LIMITS IN OCCUPATIONAL EXPOSURE TO HEAVY METALS

Members

Dr L. K. A. Derban, Chief Medical Officer, Volta River Authority, Accra, Ghana (*Vice-Chairman*)

Professor M. H. Noweir, Director, Department of Occupational Health, Institute of Public Health, University of Alexadria, Egypt

Professor W. O. Phoon, Head, Department of Social Medicine and Public Health, University of Singapore, Singapore

Dr P. K. Suma'mur, Director, Institute of Occupational Health, Department of Manpower, Djakarta, Indonesia

Professor I. P. Ulanova, Institute of Industrial Hygiene and Occupational Diseases, Moscow, USSR

Professor R. L. Zielhuis, Coronel Laboratory, University of Amsterdam, Netherlands (*Chairman*)

Representatives of other organizations

United Nations Environment Programme

Dr S. Milad, Scientific Officer, UNEP/IRPTC, Nairobi, Kenya

Mr S. K. Satkunananthan, Chief, Scientific Programme, UNEP/IRPCT, Geneva, Switzerland

International Labour Organization

Dr D. Djordjevic, Occupational Safety and Health Branch, ILO, Geneva, Switzerland

Commission of the European Communities

Dr A. Berlin, Health and Safety Directorate, Luxembourg

Permanent Commission and International Association on Occupational Health

Professor R. Truhaut, Department of Toxicology and Industrial Hygiene, University of Paris, France

United States National Institute for Occupational Safety and Health

Dr K. Bridbord, Director of the Office of Extramural Coordination and Special Projects, Rockville, MD, USA

Secretariat

Dr M. A. El-Batawi, Chief Medical Officer, Office of Occupational Health, WHO, Geneva, Switzerland (*Secretary*)

Dr E. J. Fairchild, Scientist, Office of Occupational Health, WHO, Geneva, Switzerland

Professor L. Friberg, Head, Department of Environmental Hygiene, Karolinska Institute, Stockholm, Sweden (*Temporary Adviser*)

Dr S. Hernberg, Scientific Director, Institute of Occupational Health, Helsinki, Finland (*Temporary Adviser*)

Professor F. Kaloyanova, Director, Institute of Hygiene and Occupational Health, Academy of Medical Sciences, Sofia, Bulgaria (*Temporary Adviser*)

Professor R. Lauwerys, Head, Industrial and Medical Toxicology Unit, Louvain University, Brussels, Belgium (*Temporary Adviser*)

Professor M. Saric, Director, Institute for Medical Research and Occupational Health, Zagreb, Yugoslavia (*Temporary Adviser*)

RECOMMENDED HEALTH-BASED LIMITS IN OCCUPATIONAL EXPOSURE TO HEAVY METALS

Report of a WHO Study Group

A WHO Study Group on Recommended Health-Based Limits in Occupational Exposure to Heavy Metals met in Geneva from 5 to 11 June 1979. Dr M. A. El-Batawi, Chief Medical Officer, Office of Occupational Health, opened the meeting on behalf of the Director-General. He said that the purpose of the Study Group was to make recommendations on health-based limits in occupational exposure to cadmium, lead, manganese, and mercury. A further purpose was to standardize methods of applying existing information in making decisions on what could be considered safe levels of exposure to toxic substances and to establish a policy that could be further elaborated in countries for undertaking this work. He also said that in the WHO programme on internationally recommended health-based limits in occupational exposure emphasis is laid on epidemiological studies among industrial workers and consideration is given secondarily to experimental studies on animal models.

1. INTRODUCTION

This report on heavy metals is the first to be published in response to the 1977 Executive Board resolution EB60.R2 (*1*), which urged WHO to implement as soon as possible a programme on internationally recommended health-based limits in occupational exposure to toxic substances.

It should be noted that the exposure limits recommended in this report are not in any manner binding on any administrative authority in any country; they are only for the purpose of reference and authorities concerned should set their own operational exposure limits.

1.1 Current status of exposure limits

One major goal of occupational health programmes is to prevent health impairment from exposure to harmful agents in the working environment; one of the most important ways of accomplishing this is to establish some "permissible level" of a harmful substance in the workroom air—i.e., a level at which that substance does not cause adverse health effects during the lifetime of the worker.

In 1976 a WHO Expert Committee on Methods used in Establishing Permissible Levels in Occupational Exposure to Harmful Agents (*2*) defined the term "permissible level" as a quantitative hygienic standard for a level to be considered safe, expressed as a concentration with a defined average time.

The present Study Group preferred to use the term "recommended health-based occupational exposure limit" in place of "permissible level". This term is in accordance with the International Convention (148) adopted by the International Labour Conference in 1977 (*3*). (This Convention uses the term "occupational exposure limit" for acceptable concentrations of harmful agents in workroom air.) The term "recommended health-based occupational exposure limit" represents, in the opinion of the Study Group, levels of harmful substances, in workroom air at which there is no significant risk of adverse health effects (see page 11); this does not take into account technological and economic considerations and therefore should be distinguished from operational occupational exposure limits.

In addition to the recommended health-based exposure limits for workroom air, this report also proposes corresponding recommended health-based biological limits (i.e., the no-adverse-effect level of toxic substances or their metabolites in human biological materials); these are relatively recently developed indicators of permissible exposure. In some cases, a range has been recommended instead of a definite figure. To be on the safe side, the Group set the recommended exposure and biological limits below the detected adverse-effect/response levels.

Occupational exposure limits have been set in an increasing number of countries. The two leading countries in this respect are the USA and the USSR. The exposure limits in different countries vary considerably, and often much more than one would expect on the basis of the available scientific evidence. At the meeting of the Joint ILO/WHO Committee on Occupational Health in 1968 (*4*), it was found that only 24 out of some 600 substances had approximately the same permissible limits in the USA and the USSR. There is now evidence to suggest that some convergence with regard to limits is taking place. However, there are still vast differences, which are a matter of concern for international organizations such as ILO, WHO, and the Permanent Commission and International Association for Occupational Health. These organizations have attempted to eliminate differences in occupational exposure limits in different countries in order to promote similar health standards for all industrial workers in the world.

Most discrepancies between different schools of thought on occupational exposure limits can be accounted for by different standard-setting processes, which may or may not consider technological and economic factors. While most countries base their occupational exposure limits on prevention of adverse health effects, some countries also take into account technological and economic factors.

At the international level there can be no mechanism for incorporating economic and technological factors into decision-making with respect to occupational exposure limits because these factors vary considerably between different countries. However, WHO can develop *health-based* recommendations for international consideration, leaving to national authorities the responsibility of finding ways of making use of them.

The basic objectives in establishing occupational exposure limits are very similar for the USA, the USSR, and other countries. The US Occupational Safety and Health Act of 1970 (*5*) states that exposure limits are needed in order "to ensure that no employee will suffer impaired health or functional capacities or diminished life expectancy as the result of his work experience". In the USSR, the occupational exposure limit is that level of exposure which "in the case of daily exposure at work (excluding rest days) for 8 hours, or some other duration but not totalling more than 41 hours a week, will not cause in the individual any disease or disorder from a normal state of health, detectable by current methods of investigation, during his working life or during his entire lifetime, or in any of his descendants" (*6*).

It should be noted that recently the USSR and a few other countries have included prevention of health impairment in the offspring as a criterion in determining exposure limits.

1.2 Criteria for the selection of substances of highest priority

The number of harmful industrial agents is so vast and the present knowledge about them is so limited that it is virtually impossible to set occupational exposure limits for all of them. It is therefore necessary to select certain priority substances on which there is sufficient information. The four heavy metals discussed in this report were selected on the basis of the following criteria:

— the distribution and abundance of the agent, and the frequency of exposure (or potential exposure) to it;

— the potential of the agent to cause serious functional disability; and

— the availability of reliable scientific evidence based on epidemiological and experimental studies.

The four metals also show wide discrepancies in their occupational exposure limit values in different countries.

Some of the principal occupational exposures to these metals occur as follows.

(1) *Cadmium* exposure is encountered in the primary cadmium and lead-zinc industries. Cadmium is a common constituent of alloys with low melting-points; the use of solders may expose workers to high concentrations of cadmium oxide fumes. Other sources of exposure include the metal-plating industry, the production and use of cadmium pigments and cadmium-containing plastic stabilizers, and the manufacture of nickel-cadmium batteries.

(2) *Lead* is widely used in the manufacture of batteries, in the rubber industry, and in the production of glazes, glass, and enamels. With the development of technology new uses are often found; for example, lead compounds are now being used as stabilizers in the manufacture of polymers.

(3) *Manganese* has three principal uses: (*a*) as an important constituent of certain steels and as a reducing agent in steel-making; (*b*) in the production of dry-cell batteries; and (*c*) in the chemical industry in the production of important manganese compounds such as potassium permanganate. Manganese ore mining is also a major source of occupational exposure.

(4) *Mercury* is used *inter alia* in the production of mercury vapour lamps, rectifiers, batteries, electrical switches, and relays for electronic equipment. Since 1947, the amount of mercury used in the manufacture of sodium hydroxide has increased strikingly, and this now represents the largest single use of mercury in industry. Some

inorganic and organic mercury compounds are used in agriculture as pesticides e.g., phenylmercury and methoxyethylmercury.

1.3 Periodic reappraisal

The limits recommended in this report do not represent the official policy of any international organization. They represent the conclusions reached by a group of experts after reviewing documents prepared by WHO Collaborating Centres. It is recognized that these limits have a temporary character and are subject to periodic reappraisal in the light of new evidence.

1.4 The "two-step" procedure for establishing occupational exposure limits

The WHO Expert Committee on Methods Used in Establishing Permissible Levels in Occupational Exposure to Harmful Agents (*2*) concluded that "... occupational toxicologists, physicians and hygienists have reached a broad agreement on the approaches and the methods to be used for providing the basic scientific information needed to recommend, evaluate, and revise permissible levels for occupational exposure". This was regarded as a major step towards developing international recommendations for permissible levels, but it was pointed out that "... differences exist in the way Member States translate health-based permissible levels for occupational exposure into educational, technical, compliance and enforcement measures directed towards protecting workers' health".

The establishment of occupational exposure limits requires a two-step procedure: the first step is the development of *health-based* recommended occupational exposure limits on the basis of scientific evidence judged by experts; the second step is the translation of these health-based limits into *operational* limits or standards after discussions between the government and representatives of employers and workers. The operational limits may then be enforced if necessary.

The limits recommended in this report have been agreed upon by experts from various countries, and represent the first step in the two-step procedure. It is hoped that this will facilitate policy decisions in the second step.

1.5 Relations between exposure-effect and exposure-response

In considering the effects of toxic substances in the atmosphere, the concepts of "exposure-effect" and "exposure-response" are useful. They correspond to the terms "dose-effect" and "dose-response" used in pharmacology and experimental toxicology, in which exposure doses are precisely known. However, in occupational exposure the individual doses can only be approximate since skin absorption and effects of physical activity cannot be estimated accurately. Although in the literature both terms are often used interchangeably, the Group decided that these terms should be used differently for different concepts, as used in other WHO documents (*2, 7*).

An *exposure-effect relationship* is the relationship between quantified exposure and the quantitative severity of a health effect in an individual or group. An *exposure-response relationship* is the relationship between quantified exposure and the percentage of individuals with an effect of specified severity. Examples of exposure-effect relationships are given in the chapter on lead, which relates the accumulation of porphyrins in erythrocytes or increased excretion of ALA (δ-aminolevulinic acid) in urine to the level of lead in blood. On the other hand, an exposure-response relationship is illustrated by data in Table 6, which shows the percentage of subjects with scores $\geqslant 1$ or $\geqslant 2$, i.e., the response at various exposure levels.

It is important to distinguish between the two concepts. The exposure-effect relationship exhibits an *average* effect in all individuals at the same exposure levels, thereby suggesting that all individuals can be considered to be more or less homogeneous. An exposure-response relationship, however, takes into account the variation in susceptibility within a group of individuals; it indicates the proportion of persons affected. From the above it would appear that protection of workers should be primarily based on exposure-response data. In other words, the aim is not to protect the fictitious "average" worker, but those who are most sensitive to exposure.

With increasing exposure the severity and the number of adverse effects increase; and with decreasing exposure, a limit is reached below which no adverse effects or responses are observed. These are the *no-adverse-effect* or *no-adverse-response* levels. However, since these levels are based on observations of a limited number of subjects, one can never be sure that other subjects under similar conditions would not be affected. Thus, the concept of no-*detected*-adverse-effect/response level was created and the Study Group based its recommendations on this concept.

The exposure-response relationship demonstrates that at low levels of exposure not all subjects show a response, but the probability of response and the number of different effects increase with exposure. It is important to realize that the recommended exposure limits indicate only the possibility that an individual will be adversely affected, and the fact that a person may be exposed to a certain level of a toxic substance does not necessarily mean that he will suffer from a disease. Therefore, there limits should not be used indiscriminately for clinical diagnosis.

1.6 Adverse effects

The Study Group on Early Detection of Health Impairment in Occupational Exposure to Health Hazards (*7*) pointed out that "a statistically significant effect as such is not the same as impairment of health". Furthermore, "what is regarded as unacceptable (not permissible) is a matter of interpretation and ultimately of choice".

Non-adverse effects are defined by the National Academy of Sciences (*8*), USA, as:

(1) changes that occur with continued exposure and do not result in impairment of functional capacity or the ability to compensate for additional stress;

(2) changes that are reversible following cessation of exposure if such changes occur without detectable decrements in the ability of the organism to maintain homeostatis, and;

(3) changes that do not enhance the susceptibility of the organism to the deleterious effects of other environmental influences—whether chemical, physical, microbiological, or social.

In a recent WHO publication (*9*), an *adverse* or "abnormal" effect has been defined in terms of a measurement that is outside the normal range. Sanockij (*10*) considered departures from "normal" values as adverse effects (1) if the observed changes are statistically ($P < 0.05$) outside the limits of generally accepted "normal" values, (2) if within the normal range the changes persist for a considerable time after the cessation of exposure, or (3) if departures from the generally accepted "normal" values become manifest under functional or biochemical stress.

Criteria differentiating between adverse and non-adverse effects should not be based on overt pathology. Many toxicologists have proposed a number of criteria based on metabolic and biochemical changes that occur in the body as a result of exposure to toxic substances. These include: decreased efficiency of metabolic processes and abnormal excretion of certain substances, inhibition of key enzymes, increased concentration of natural substrates induced by enzyme inhibition and/or decreased ability to metabolize specific substrates, and change in relative activities of different enzyme systems.

The limits recommended in this report are expected to prevent not only overt disease, but also adverse health effects in workers exposed to harmful agents throughout their working life and in their offspring. There was a general consensus within the Group regarding the intensities and effects to be considered as "adverse", even though the clinical significance of many "adverse" effects has not been unequivocally established.

The Group considered the following types of effects as adverse:

— effects that indicate early stages of clinical disease;
— effects that are not readily reversible, and indicate a decrement in the body's ability to maintain homeostasis;
— effects that enhance the susceptibility of the individual to deleterious effects of other environmental influences;
— effects that cause relevant measurements to be outside the "normal" range, if they are considered as an early indication of decreased functional capacity; and
— effects that indicate important metabolic and biochemical changes.

For details on early detection of health impairment due to occupational exposure, see references *2, 7*, and *9*.

Assessment of adverse effects of occupational exposure must be based on a comprehensive set of physiological impairment data. Assessments based on such data are vital in making decisions on health-based occupational exposure limits. Differentiation between "non-adverse" and "adverse" effects requires considerable knowledge of reversible changes, which involve subtle departures from "normal" physiology and tissue morphology; the ability of the body to adapt to subtle changes should be accounted for in such differentiations.

1.7 Assessment of exposure

The concentrations of chemical substances in air, used as indications of occupational exposure, are often measured by *area (static) sampling*, in which samples are taken from specific points in the workplace. Very few data are based on *personal sampling*, in which estimates are made of the concentrations of harmful agents in the air inhaled by the worker. Despite recent advances in instruments and techniques of personal sampling, the data on personal exposure are still not sufficiently reliable because oral and cutaneous intake of harmful agents is usually not accounted for in them. The four metals included in this report may enter the body through contaminated food and beverages, and smoking has a relatively large impact on cadmium intake. These aspects are particularly important because the worker's family members may be affected by his contaminated work clothes.

Since total exposure is often not reflected in measurement of air levels alone, this report gives considerable attention to the *biological* assessment of exposure, and wherever possible, the health-based biological limits have also been stated.

Until a few years ago, data on occupational exposure often referred to group average levels (average level of a harmful substance or its metabolites in a group of subjects). For certain metals (e.g., mercury) a relation between levels of metal in air and group average levels has been observed. Nowadays, however, it is more common to refer to individual exposure levels since biological limits to assess individual exposure have been established. It is believed that the health of the workers can be better protected by relying on individual health-based biological limits than on group average levels. One of the drawbacks of measuring levels of metals in air is that they indicate only the respiratory exposure, and, to a certain extent, the cleanliness of the workplace while the impact of oral and/or cutaneous intake can be seen only by determining the biological level. In addition, the non-occupational background exposure is also expressed in the biological level. For the above reasons, wherever possible, emphasis is laid on individual health-based biological limits.

Occupational exposure limits based on levels of harmful substances in air are very useful to industrial hygienists and technologists in planning and designing industrial complexes with safer and more hygienic working environments. However, personal hygiene must be respected and the oral intake of harmful substances from occupational and non-occupational sources must be prevented by such measures as providing wash-hand basins at the workplace and facilities for changing contaminated clothes frequently.

1.8 Analytical problems

The exposure-effect and exposure-response relationships presented in the following four sections are based on studies of the working environment or analysis of blood and urine samples, or on both. While there are analytical problems in both cases, they are more acute in the latter, which requires the measurement of metal levels in blood or urine ranging from 1 to > 1000 μg/l.

In recent years, much work has been done on interlaboratory comparison studies with the aim of ensuring the reliability of research data. These studies have demonstrated that data can be considered reliable only if they are reproducible within a standard deviation. The methods used to obtain them must be acceptable in sensitivity and precision.

One of the basic problems in the analysis of workroom air is the accurate estimation of true exposure. In measuring exposure one has to consider not only the general concentration of metal in the air, but, in the case of aerosols, the particle size and distribution. Attention should also be paid to "respirable particles" whose concentration may vary from time to time and place to place. The solubility of the harmful agent should also be taken into account.

The interpretation of assessed exposure is likely to be indirectly influenced by the sampling and analytical methods adopted; attention should be given to this fact when exposure data are being interpreted.

1.9 Application of occupational exposure limits

1.9.1 *Specific purposes*

The policy decisions on setting operational exposure limits, which constitute the second step in the two-step procedure (see page 11), may differ in various countries. Operational exposure limits are used in various way:

— in planning new plants making use of existing technology or developing improved technology that better protects workers' health;
— in designing ventilation systems and new machines when reconstructing existing plants;
— in applying sanitary control measures; and
— in assessing the health risk for individual workers.

1.9.2 *Health-based occupational exposure limits and developing countries*

The differences in existing exposure limits have led to uncertainty in the choice of values to be implemented in many developing countries, and may have been instrumental in delaying the protection of workers from harmful agents. The establishment of internationally accepted occupational exposure limits should ameliorate this situation.

Industrial workers are particularly at risk of health impairment because along with the risk of being affected by the diseases prevailing in the community they are subject to occupational hazards. When for technical or economic reasons, it is not possible to introduce preventive measures, such as replacing toxic industrial substances by harmless ones or mechanizing industrial processes in which harmful agents are used, it is essential to establish occupational exposure limits.

It should be recognized that conditions in the developing countries sometimes impose constraints on the application of exposure limits and may influence the authorities with regard to their adoption. Health planners point to the need to protect workers from harmful industrial agents owing to their greater vulnerability, but other planners argue that stringent legislation may increase production costs considerably. To a large extent this situation is also observed in the developed countries.

Workers suffering from parasitic diseases and malnutrition in developing countries should not, as a policy, be exposed to toxic agents in the working environment, as this would only aggravate their health problems. Large industries and multinational companies in the developing countries can afford pre-employment medical screening of workers and the treatment of those found to be suffering from a disease. The discovery of a disease state in pre-employment medical examination should not be used as a means of excluding workers from employment but should be viewed as an opportunity to treat them and ascertain that they start their work as healthy people.

1.9.3 *Operational versus recommended health-based occupational exposure limits*

The health-based occupational exposure limits recommended in this report are more rigorous than the existing operational occupational exposure limits in most countries. They allow the authorities to estimate the degree of risk they will be taking if they adopt operational exposure limits that exceed the proposed health-based occupational exposure limits. This does not always necessarily imply that there will be an increase in the incidence of overt occupational diseases, but it can involve the occurrence of adverse effects in a number of workers and/or their offspring. If the implementation of recommended health-based occupational exposure limits as operational occupational exposure limits is not feasible at short notice, the health-based occupational exposure limits should be regarded as goals to be achieved as soon as possible.

1.9.4 *Occupational exposure limits and duration of exposure*

For exposures that do not exert immediate effects and for agents that have a relatively short biological half-life, occupational exposure limits can be expressed as time-weighted-average limits per workshift. Because industrial exposures fluctuate, it may be necessary to introduce a maximum limit that must not be exceeded in the application of operational occupational exposure limits, but this aspect is not discussed in this report. If the fluctuations do not differ too greatly from the time-weighted-average level there is no immediate health risk.

For certain health effects, which are due to direct local exposure of the respiratory tract, as in the case of cadmium, mercury, and manganese, an extra health-based short-term exposure limit may be necessary to protect workers againsts the risk of incidental high exposure (e.g., in the welding of cadmium). These short-term limits should take into account the proposed time-weighted-average limit.

1.9.5 *Interpretation of the recommended health-based occupational exposure limits*

The health-based occupational exposure limits recommended in this report should not be interpreted as absolute levels above which adverse effects will occur and below which adverse effects will not occur. Not only is there too much variation in individual susceptibility, but there are also inaccuracies in the analysis of biological specimens. When interpreting the results of biological tests on workers' blood or urine samples, one should particularly note the trend of the observed biological levels over a period of time and also see if there are any early indicators of health impairment.

Individual exposure should be judged on the basis of not just one indicator but, whenever possible, of *all* recommended indicators. A discrepancy between the levels in air and in blood, for example, may point to exposure not directly due to inhalation of workroom air. Moreover, in some regions the non-occupational oral exposure to metals such as lead and mercury may be relatively high owing to atmospheric pollution or food or water contamination. In such cases blood samples will continuously show high exposure and therefore health-based occupational exposure limits will have to be lower than those recommended in this report.

1.10 Research possibilities

The following four sections on cadmium, lead, manganese and mercury contain an account of ongoing and future research. The aim of this is to emphasize the topics that are currently being studied with regard to health-based occupational exposure limits. Also included are recommendations for further research to fill gaps in knowledge.

It has been stressed that occupational exposure limits should protect the health not only of the workers, but also of their families. The occurrence of harmful effects on reproductive function in males and females (gonadotoxic effects), on the development of the fetus (embryotoxic effects), on the production of malformations or deviations from normal structure in offspring (teratogenic effects), and on the postnatal development and health of offspring of male or female workers needs to be studied urgently. The same may be said for carcinogenic effects, particularly when cadmium is present in the environment. Effects related to behaviour and dysfunction of the nervous system also warrant research. The section on manganese is an example which clearly demonstrates that there is a dearth of pertinent data on exposure and on various types of responses.

Many of the effects observed are not specific to the exposure in question; the appraisal of exposure-response relationships, therefore, demands elaborate research activities, and there is an urgent need to study these effects.

Experts in industrial health and occupational toxicology are invited to make available to WHO any additional relevant data they may have on exposure-effect and exposure-response relationships in occupational exposure to cadmium, lead, manganese, and mercury. Such information will be considered by WHO in its periodic re-evaluation of recommended health-based occupational exposure limits, as this becomes necessary.

REFERENCES

1. WHO Handbook of Resolutions and Decisions of the World Health Assembly and the Executive Board. Vol. II (3rd ed.), 1979, p. 38.
2. WHO Technical Report Series, No. 601, 1977.
3. INTERNATIONAL LABOUR ORGANISATION. Conventions and recommendations. ILO Document No. D 15, Geneva, 1979.
4. WHO Technical Report Series, No. 415, 1969.
5. UNITED STATES CONGRESS. *Occupational Safety and Health Act of 1970.* Washington, DC, US Government Printing Office (Public Law 91-596 Ninety-first Congress, S. 2193, December 1970.
6. GOSUDARSTVENNYI STANDART SOYUZA SSR (12. 1.005–76). Moscow, State Committee for Standards, 1976.
7. WHO Technical Report Series, No. 571, 1975.
8. COMMITTEE FOR THE WORKING CONFERENCE ON PRINCIPLES OF PROTOCOLS FOR EVALUATING CHEMICALS IN THE ENVIRONMENT. *Report*, Washington, DC, National Academy of Sciences, 1975.
9. WORLD HEALTH ORGANIZATION. *Environmental health criteria 6. Principles and methods for evaluating the toxicity of chemicals. Part I*, Geneva, 1978.
10. SANOCKIJ, I. V. In: Sanockij, I. V., ed. *Metody opredelenija toksičnosti i opasnosti himičeskih veščestv.* Moscow, Medicina, 1970.

2. CADMIUM[1]

2.1 Summary of metabolism and toxicity

2.1.1 *Metabolism*

Routes of exposure. In industry, exposure to cadmium occurs mainly by the pulmonary route, but in some circumstances ingestion of cadmium dust may also be significant. Cadmium may be ingested directly from contaminated hands (mainly when workers eat or smoke at the workplace) or indirectly following the clearance of large particles deposited in the upper respiratory tract. Only particles with an aerodynamic diameter of 5 μm or less can enter the pulmonary compartment. Cigarette smoking adds to the amount of cadmium deposited in the lung. One cigarette may contain about 0.9–2.3 μg of cadmium and it is estimated that about 10% of that amount is inhaled.

Absorption. The amount of cadmium absorbed through the lungs depends on the amount retained in them (the amount deposited minus the amount eliminated by the clearance mechanisms) and also on the chemical form of the retained particles that influences their rate of solubilization in the tissues. There is a dearth of data on the importance of these factors during chronic exposure to cadmium. Theoretical calculations based on the amount of cadmium found in the body tissues of smokers suggest that about 50% of the amount of cadmium deposited in the lungs (probably as cadmium oxide) may be absorbed. Hence, it has been estimated that when most of the particles are in the respirable range, 20–30% of the amount of cadmium inhaled may be absorbed. This rate decreases with decreasing solubility of the cadmium salts—cadmium sulfide, for instance, has a tendency to be retained in the lung.

In man, the gastrointestinal absorption of cadmium is usually less than 10%. Various dietary factors, such as iron, calcium, and protein deficiency, may increase the gastrointestinal absorption rate (*1*), which may also depend on the solubility of cadmium salts at gastrointestinal pH values. Percutaneous absorption of cadmium is considered to be negligible.

---

[1] This section is largely based on the works cited in references *4*, *5*, and *8*.

Distribution. More than 70% of the cadmium circulating in the blood is found in the red blood cells. Cadmium is a cumulative toxic agent with a biological half-life of several years. The cadmium burden of the body increases with age and is greater in smokers than in non-smokers. The main sites of deposition are the liver and kidneys. The relative importance of these two sites of storage depends on the intensity and duration of exposure and on renal function. In the body tissues, cadmium is bound principally to metallothioneine, a metal-binding protein of low molecular weight; synthesis of this protein is induced by cadmium. Estimates of the body burden of non-occupationally exposed adults living in areas not excessively polluted with cadmium have ranged from 9 mg to 40 mg and the mean cadmium concentrations in renal cortex have been shown to have a range of 10–50 mg Cd/kg (wet weight). Higher values have been found in industrial workers exposed to cadmium. The kidney is a target organ in long-term exposure to cadmium (see below) and when renal dysfunction develops the cadmium level in the kidney decreases owing to higher excretion of cadmium.

Excretion. Cadmium is excreted mainly through urine and to a much lesser extent through bile, gastrointestinal secretion, sweat, saliva, hair, and nails. Transplacental transfer and secretion into milk are very low.

2.1.2 *Significance of cadmium concentration in biological materials*

Abnormal levels of cadmium in blood (Cd–B) are an indicator of recent exposure to cadmium. However, not enough data exist to establish with sufficient accuracy the relationship between cadmium exposure from workroom air and its concentration in blood. When exposure is low (e.g., in the general population) it is not yet clear to what extent the cadmium level in blood Cd–B reflects the body burden and recent exposure.

On a group basis there is a correlation between cadmium concentration in urine (Cd–U) and body burden up to a certain level of urinary excretion of cadmium. When exposure has been so intense as to saturate all the binding sites in tissues, cadmium in urine may become more an indicator of current exposure than of body burden. The concentration of cadmium in urine corresponding to the state of tissue saturation appears to lie around 10–15 μg Cd/g creatinine.

Estimation of cadmium in hair has been proposed as a measure of body burden of cadmium. However, in practice, the difficulty of distinguishing endogenous cadmium from externally deposited cadmium makes this method unreliable.

Cadmium excreted in faeces is a satisfactory indicator of daily oral intake since the amount of cadmium absorbed or excreted daily through the gastrointestinal tract represents a small and definite fraction of the intake.

Normal levels. In non-smokers who have no occupational exposure to cadmium, concentration of cadmium in whole blood is usually below 5 μg/l; smokers have a higher concentration. Cadmium concentration in urine increases with age but in persons less than 60 years old, living in non-polluted areas, the amount of cadmium excreted daily in urine rarely exceeds 2 μg.

2.1.3 *Toxic effects of cadmium*

Acute effects. In man, the principal acute manifestations induced by cadmium are gastrointestinal disturbances following ingestion, and chemical pneumonitis following inhalation of cadmium oxide fumes.

In animals, acute administration of cadmium (mainly by parenteral routes) can produce toxic effects in many parts of the body including the kidneys, liver, testes, ovaries, nervous system, pancreas, cardiovascular system, and placenta. Sarcomata at the sites of injection and teratogenic effects may also be observed.

Chronic effects. In man, the principal toxic effects resulting from long-term exposure to cadmium are renal dysfunction and lung impairment. Other reported toxic changes include bone changes, slight anaemia, and anosmia. Although parenteral administration of cadmium can induce cancer in rats at the site of injection and interstitial cell tumours in the testes, there is at present no evidence to suggest that cadmium administered orally to animals is carcinogenic. However, it should be pointed out that the studies carried out so far have been mainly on the toxicity of cadmium rather than its carcinogenicity.

Some epidemiological studies suggest that exposure to cadmium may increase the incidence of prostate (*2*) and lung cancer. A study by Lemen et al (*3*) suggests that there may even be an increase in the risk of cancer of the respiratory tract. However, other data suggest that the risk of cancer from cadmium is low and that the results of the above-mentioned studies reflect a very high occupational exposure in the past. In the opinion of the Study Group the epidemiological studies on carcinogenicity of cadmium are not conclusive because most of them involved only a small number of workers. Furthermore, some investigators were not able to rule out the possible role of other agents (e.g., tobacco smoke) in their data. For these reasons the possible carcinogenicity of cadmium cannot be considered in deriving health-based occupational exposure limits.

Conflicting results have been reported regarding the occurrence of chromosomal anomalies in cultured lymphocytes of workers exposed to cadmium. Some studies (mainly post-mortem studies) on groups of the general population have implicated cadmium as an etiological agent in the development of cardiovascular disease, particularly hypertension. Such a relationship has not been established in cadmium workers.

In animals, repeated administration of cadmium can cause liver dysfunction, and in some strains of rats it has a slight hypertensive effect. In long-term occupational exposure to cadmium the kidney is usually the critical organ but in some individuals the lungs may be affected first.

2.2 Relationship between exposure and health effects

2.2.1 *Choice between biological and environmental exposure indicators*

The choice between biological and environmental exposure indicators depends on the target organ in question. Since acute and chronic lung disturbances induced by cadmium are most probably the result of a local toxic action, their intensity must be related to external exposure factors such as concentration in air and duration of exposure. It is also evident that the acute effect of cadmium on the gastrointestinal tract is directly related to the amount of cadmium ingested.

Since the chronic systemic effect on the kidney results from a progressive accumulation of the metal in the organ and long-term exposure to cadmium in industry occurs by inhalation and ingestion, air analysis is the most useful method of evaluating the risk to the lungs. This method may also be useful in the identification of the sources of emission and in the evaluation of the efficiency of preventive methods. However, the systemic chronic effects of cadmium are better evaluated by complementary measurements of the cadmium concentrations in blood and urine which reflect mainly recent exposure and body burden respectively.

2.2.2 *Analytical problems and their impact on the interpretation of published data*

There are three main difficulties in the analysis of the results reported in the literature. Few papers give any information on the accuracy of the methods used for cadmium determination in biological material. This is a very serious limitation, since it is known that cadmium analysis in blood and in urine is difficult and may involve many analytical errors (contamination, loss during sample pretreatment, etc.). The use of inaccurate methods may partly explain the large discrepancies among "normal" values reported by different authors.

The second difficulty concerns the different sensitivities of the methods used for detecting a biological effect of cadmium on the kidneys (e.g., total proteinuria and specific protein concentration in urine).

Finally, the data on the concentration of airborne cadmium at the workplace are very fragmentary (e.g., limited sampling time, no information on particle size), and this leads to many inaccuracies in the estimation of environmental exposure.

2.2.3 *Exposure-effect relationship*

Chemical pneumonitis manifested by dyspnoea, cough, sputum production, chest discomfort, and impairment of lung function has been reported in workers exposed to freshly generated cadmium oxide fumes for several hours at concentrations above 0.5 mg cadmium/m^3 (*4*). Delayed fatal pulmonary oedema may occur if the concentration of cadmium oxide fumes exceeds 5 mg cadmium/m^3 for 8 hours. Cadmium dust is less acutely toxic to the lungs than cadmium oxide fumes. It has been estimated that an 8 h exposure to a "respirable" cadmium dust concentration above 3 mg/m^3 may cause acute respiratory symptoms.

Administration of a single oral dose of 10 mg of cadmium may lead to gastrointestinal disturbances (nausea, vomiting) but the acute oral lethal dose for an adult is probably above 350 mg.

Long-term excessive exposure to cadmium has been reported to induce first either proteinuria or lung disturbance. Regarding the latter, it is not clear from the literature whether it results from long-term exposure above the critical airborne cadmium concentration or from several episodes of subacute exposure leading to permanent lung damage. This situation has mainly resulted from the fact that in early investigations sufficient consideration was not always given to the effects of tobacco smoke on the lungs.

Various types of lung dysfunction have been described (e.g., emphysema, chronic obstructive pulmonary disease, and lung fibrosis). However, in the absence of subacute episodes of chemical pneumonitis, signs of renal dysfunction will probably appear before signs of respiratory impairment.

Signs of renal disturbance found in workers exposed to cadmium include: increased total proteinuria; increased renal clearance of specific proteins (β2-microglobulin, retinol-binding protein, albumin, transferrin, IgG); increased plasma urea and creatinine; and perturbation of some functional tests such as creatinine clearance, inulin clearance, uric acid clearance, acid load test, test for estimating the tubular reabsorption rate of phosphorus and calcium, and urine concentrating ability test. A high prevalence of renal stones has also been found in some workers suffering from cadmium poisoning.

Proteinuria, when measured by a sufficiently sensitive and quantitative technique, precedes other signs of renal damage. It is usually due to tubular damage, which results in increased excretion of proteins of low molecular weight, but in some cases it may result from early glomerular dysfunction, in which case proteins of high molecular weight are excreted.

Disturbance of calcium metabolism and bone lesions (osteomalacia and/or osteoporosis) have been reported mainly in workers exposed to cadmium over a very long period of time. In a few cases, however, bone lesions have also been observed in workers exposed for less than five years. The pathogenesis of the condition has not yet been fully elucidated but it has been seen that a majority of workers who develop such lesions already have signs of renal dysfunction.

Symptoms of central nervous system dysfunction have also been reported in workers in an accumulator factory but no data on the relationship with the intensity of exposure are available.

2.2.4 *Critical adverse effects*

The available data suggest that in workers chronically exposed to cadmium the critical effects are kidney and lung dysfunction. This conclusion may have to be revised when further data on the carcinogenic effect of cadmium in man become available.

Lung disturbances can only be prevented by keeping the concentration of airborne cadmium dust or fumes below a certain critical level. However, in order to prevent renal damage, the absorption of cadmium from all routes (pulmonary and oral) must be taken into consideration. The impact of combined exposure can be evaluated only by determining the level of cadmium in blood and urine. Therefore, the concentration of cadmium in these biological materials should be kept below the level at which dysfunction of vital organs occurs.

2.2.5 *Exposure-response relationships for the critical effects*

Effect on the lungs (Table 1). Only a limited number of investigations have been carried out on the lung function of workers with long-term exposure to cadmium. Unfortunately, these studies were not based on identical tests. Many epidemiological studies suggest that in the absence of acute episodes of high exposure to cadmium, the changes induced by cadmium inhalation in the lungs are usually mild. It should be stressed, however, that a majority of epidemiological studies have been performed on active workers, and this may have introduced a bias in the sense that only "resistant" workers may have been selected.

The paucity of data on the exposure-response relationship for the action of cadmium on the lungs permits only a tentative proposal for a long-term no-adverse-response level. It seems that to prevent any deleterious effect on the respiratory system the time-weighted-average exposure to cadmium oxide fumes or to respirable cadmium dust should not exceed 20 μg Cd/m^3 (duration of exposure: 40 h per week during the whole working life). It would not be wise to base the occupational exposure limit on total dust concentration. Not enough data exist to distinguish between different chemical forms of cadmium dust.

Effect on the kidneys. In long-term low-level occupational exposure to cadmium, the kidneys rather than the lungs are more frequently the critical organs. The first sign of kidney dysfunction is usually increased proteinuria. Cases of renal disturbance, sometimes associated with bone disease (itai-itai disease), have also been found in the general population living in cadmium-contaminated areas of Japan. In 1977 a WHO task group (*5*) concluded that daily intake of 300–480 μg of cadmium could increase the prevalence of proteinuria.

Table 1. Clinical studies on the chronic effect of cadmium on the lungs

Author and reference	Number and sex of workers	Conditions of exposure	Observations
Princi, F. *J. Industr. Hyg.*, **29**: 315 (1947)	20 males	cadmium oxide fumes and dust; 40–1440 μg/m³; 0.5–22 years exposure (mean = 8 years)	no chest X-ray change
Hardy, H. L. & Skinner, J. B. *J. Industr. Hyg.*, **29**: 321 (1947)		cadmium oxide fumes and dust; mean = 100 μg/m³; 4–8 years exposure	no chest X-ray change
Friberg, L. *J. Industr. Hyg.*, **30**: 32 (1948)	43 males	cadmium oxide dust in alkaline accumulator factory; 3–15 mg/m³; 9–34 years exposure (mean = 20 years)	⅓ of subjects showed disturbance of lung function tests
Friberg, L. *Acta Med. Scand.*, **138**: Suppl. (1950)	15 males	less than 4 years in the same factory as above	no disturbance of lung function tests
Baader, E. W. *Dtsch. Med. Wschr.*, **76**: 484 (1951)	8 males	cadmium oxide dust in alkaline accumulator factory; 8–19 years exposure	6 subjects with chest X-ray suggestive of emphysema
Bonnel, J. A. *Brit. J. Ind. Med.*, **12**: 181 (1955) Kazantzis, G. *Brit. J. Ind. Med.*, **13**: 30 (1956) Buxton, R. *Brit. J. Ind. Med.*, **13**: 36 (1956) Bonnel, J. A. et al. *Brit. J. Ind. Med.*, **16**: 135 (1959)	100 males	cadmium oxide fumes; 1st factory = 13–89 μg/m³ (mean = 40–50 μg/m³) 2nd factory = 1–270 μg/m³ (mean = 132 μg/m³) (but working conditions worse in the past)	21 subjects showed disturbance of lung function tests and chest X-ray suggestive of emphysema
Kazantzis, G. et al. *Quart. J. Med.*, **32**: 165 (1963)	12 males	cadmium pigment factory; 0.5–31 years exposure (exposure to cadmium oxide dust and fumes in earlier years)	no abnormal chest X-ray, among 6 workers with more than 25 years exposure three had: — exertional dyspnoea and — disturbance of lung function tests
Potts, C. L. *Ann. Occup. Hyg.*, **8**: 1 (1965)	70 males	cadmium oxide dust in alkaline accumulator factory; up to 1949: 0.6–23.6 mg/m³ since 1950: <0.5 mg/m³; >10–40 years exposure	emphysema and chronic bronchitis in 4 workers

Table 1 (continued)

Author and reference	Number and sex of workers	Conditions of exposure	Observations
Suzuki, S. et al. *Industr. Hlth,* **3**: 73 (1965)	19 males	cadmium stearate dust; 30–690 μg/m³; on the average 3.3 years exposure	no disturbance of lung function tests
Tsuchiya, K. *Arch. Environ. Hlth,* **14**: 875 (1967)	13 males	smelting alloys of silver and cadmium; 68–241 μg/m³ (mean = 126 μg/m³); 0.75–12 years exposure	no abnormal chest X-ray
Adams, R. G. et al. *Quart. J. Med.,* **38**: 425 (1969)	27 males	cadmium oxide dust in alkaline accumulator factory since 1957; 1st area, 0.3–5 mg/m³ 2nd area, 0.1–1.0 mg/m³ 3rd area, 0.05–0.2 mg/m³; 5–44 years exposure	5 showed disturbance of lung function tests
Teculescu, D. B. & Stanescu, D. C. *Arch. Arbeit. Med.,* **26**: 335 (1970)	11 males	cadmium oxide fumes and dust; 1.21–2.7 mg/m³; 7–11 years exposure	no disturbance of lung function tests
Lauwerys et al. *Arch. Environ. Health,* **28**: 145 (1974)	26 females	mixture cadmium oxide and cadmium sulfide dust; 6.8–18.6 μg/m³ (respirable <4 μg/m³); average duration of exposure 4.4 years	no disturbance of lung function tests
Materne et al. *Cah. Med. Travail,* **12**: 1 (1975) (the two papers being considered jointly)	6 females 21 males	cadmium oxide dust in alkaline accumulator factory; 1–465 μg/m³ (respirable <90 μg/m³); average duration of exposure—females 2.8 years, males 8 years	very slight disturbance of lung function tests in males (mean values different from control but no increased prevalence of abnormal tests)
	25 males	cadmium oxide fumes and dust; 3.7–356 μg/m³, (respirable <21 μg/m³); average duration of exposure 24.5 years	slight disturbance of lung function test in 6 workers
	66 males	cadmium oxide dust and fumes; 36–25 600 μg/m³ (respirable mean = 29 μg/m³); average duration of exposure: 8 years	slight disturbance of lung function tests

Table 1 (continued)

Author and reference	Number and sex of workers	Conditions of exposure	Observations
Smith, T. J. et al. *Amer. Rev. Resp. Dis.,* **114**: 161 (1976)	17 males	cadmium oxide dust and fumes >50 μg/m^3 (up to 20 mg/m^3); average duration of exposure: 26.4 years	disturbance of lung function tests, 5 workers showed fibrosis on X-ray plates
Stanescu, D. C. et al. *Scand. J. Resp. Dis.,* **68**: 289 (1978)	18 males	cadmium oxide dust and fumes, duration of exposure >20 years (average = 32 years)	slight disturbance of lung function tests, no chest X-ray change
Gill, P. F. *Metal Bulletin Limited,* London, 1978, p. 207 (Proceedings of the First International Cadmium Conference)	34 males	cadmium oxide dust and fumes; 0.15–31 μg/m^3; 3–37 years of exposure	no disturbance of lung function tests, in a high exposure group (Cd in air: 19–31 μg/m^3) increased prevalence of exertional dyspnoea (22/34 as against 11/34 in controls) increased prevalence of respiratory infection

Data on exposure to airborne cadmium and the prevalence of proteinuria are given in Table 2. The results suggest that, even assuming a negligible intake of cadmium by ingestion, the level of cadmium dust or cadmium oxide fumes should be kept well below 20 μg/m^3 in order to prevent an increased prevalence of proteinuria in long-term cadmium exposure. However, as mentioned above, such an estimate may be misleading, because it does not take into account all factors influencing the dose of cadmium absorbed by the worker (e.g., solubility of cadmium particles and particularly direct oral absorption). Studies based on the evaluation of cadmium levels in blood and urine will yield more accurate data on the response to exposure.

The WHO task group, on the basis of evidence from autopsy studies, also concluded that the critical level of cadmium in the renal cortex for the appearance of proteinuria is between 100 and 300 mg/kg wet weight (*5*). However, the results of a recent investigation, during which the cadmium level in the kidneys of 309 cadmium workers was measured by neutron activation with concomitant evaluation of kidney function, show that the critical concentration of cadmium in the renal cortex lies between 200–250 mg/kg wet

Table 2. Prevalence of proteinuria in cadmium workers

Cadmium compounds	Estimated concentrations in the air in μg/m³	Exposure range in years	Number of examinees	Prevalence of proteinuria (%)	Method of detecting proteinuria	References
Cadmium oxide fumes	40–50	control 1–9 >9	60 37 63	2 24 46	sulfosalicylic acid method and tungstate method	King, E. *Brit. J. Industr. Med.*, **12**: 198–205 (1955) Bonnel, J. A. *Brit. J. Industr. Med.*, **12**: 181–195 (1955) Bonnel, J. A. et al. *Brit. J. Industr. Med.*, **16**: 135–147 (1959)
	64–241 [a] 123 (time-weighted average)	control <1 1–4 5–12	11 4 4 4	0 0 50 100	tungstate method >100 mg/l	Tsuchiya, K. *Arch. Environ. Hlth*, **14**: 875–880 (1967)
Cadmium oxide dust	400–15 000	1–4 9–15 16–22 23–34	15 12 17 14	0 33 41 64	nitric acid method ("Hellers test", positive in more than half the tests)	Friberg, L. *Acta Med. Scand.*, **138**: Suppl., p. 240 (1950)
	130–1170	1–12	26	27	not stated	Horstowa, H. et al. *Medycyna Prady*, **17**: 13–25 (1966)
	500 300–5000 50–1000 100–1000 50–200	5–44 5–44 5–44 5–44 5–44	116 13 43 40 20	28 15 51 20 0	sulfosalicylic acid method	Adams, R. G. et al. *Quart. J. Med.*, **38**: 425–442 (1960)

Table 2 (continued)

Cadmium compounds	Estimated concentrations in the air in $\mu g/m^3$	Exposure range in years	Number of examinees	Prevalence of proteinuria (%)	Method of detecting proteinuria	References
		control	31	0	abnormal electro-phoretic pattern as defined by the authors	Lauwerys, R. et al. *Arch. Environ. Hlth,* **28**: 145–148 (1974)
	31 [a] (1.4) [b]	1–12 (4) [c]	31	0		
		control	27	4		
	134 (88) [b]	0.6–19.7 (9) [c]	27	15		
		control	22	0		
	66 (21) [b]	21–40 (28) [c]	22	68		
	0	control	87	3.4	radioimmunoassay	Kjellström, T. et al. *Environ. Res.,* **13**: 303–317 (1977)
	50	0–3	50	6.0		
	50	3–6	30	6.6		
	50	6–12	21	19.0		
Cadmium stearate dust	30–690	control	24	17	qualitative trichloroacetic acid method	Suzuki, S. et al. *Industr. Hlth,* **3**: 73–85 (1965)
		3	19	58		
Cadmium sulfide dust	114 [d]	<1–5	12	17	electrophoresis	Harada, A. & Shibutanni, E. *Kankyo Hoken Report No. 24,* Tokyo, Japan Public Health Association, 1973, pp. 16–22 (in Japanese)
		5–21	7	100		
		<1–5	12	8	quantitative tri-chloroacetic acid method	
		5–21	7	43		

[a] Measured in the breathing zone.

[b] Respirable fraction (particle size <5 µm in diameter).

[c] The numbers in parenthesis are averages.

[d] Calculated average exposure for the worker with the most pronounced effect.

weight (*6*). The correlation between cadmium in kidney and cadmium in urine suggests that the "critical level" (minimum-adverse-effect level) of the latter is 10–15 μg Cd/g urinary creatinine. This estimate is in agreement with some recent clinical observations (*7*) which show that when, in cadmium-exposed workers, the Cd–U level exceeded 10 μg/g urinary creatinine, the prevalence of increased urinary excretion of various proteins (β_2-microglobulin, albumin, transferrin, IgG, etc.) was greater than that in matched control groups. The biological limits recommended in this report are based on the above mentioned clinical studies, and they correspond to the biological thresholds derived from the application of Friberg's model for cadmium metabolism in man (*8*).

The relationship between exposure level and cadmium concentration in blood is not yet sufficiently understood to derive a biological limit for blood with satisfactory precision. However, the Study Group agreed that the value of 10 μg Cd/l of whole blood suggested by Bernard et al. (*7*) should be accepted as a tentative no-adverse-effect level.

2.2.6 *Conclusions*

Exposure of less than 1 hour to cadmium oxide fumes and dust at a concentration not exceeding 250 μg Cd/m^3 would not lead to the occurrence of any lung reaction in workers with normal lung function. To prevent any pulmonary effect of cadmium in long-term occupational exposure, the time-weighted average airborne concentration of cadmium oxide fumes or respirable dust should not exceed 20 μg Cd/m^3 (for a weekly exposure of 40 hours during the whole working life).

In order to prevent renal dysfunction, the amount of cadmium in the renal cortex should be kept below 200 mg/kg wet weight. This level corresponds to a urinary excretion of 10 μg Cd/g creatinine (confirmed by repeated determinations every few weeks). For blood, a value of 10 μg Cd/l of whole blood is proposed as a no-adverse-effect level for long-term exposure. In addition, the concentration of respirable dust should be well below 20 μg Cd/m^3.

2.3 Research possibilities

Further studies should be carried out on the relationships between cadmium exposure, body burden, and cadmium levels in blood. The validity of the proposed health-based biological limit for cadmium in urine should also be tested by examining additional groups of workers exposed to cadmium. These results justify continuous efforts to improve the comparability of the methods used for the determination of cadmium in biological materials.

A better estimation of the no-adverse-effect level of cadmium in air with respect to effects on the lungs requires further prospective and retrospective epidemiological studies in which the exposure (concentration and duration) is more accurately known.

The important and urgent question regarding the carcinogenicity of cadmium for man justifies additional large-scale retrospective studies on groups of workers with different intensities and types of exposure. Their objectives should be twofold to conclude whether in some conditions cadmium acts as a human carcinogen, and to evaluate whether exposure conditions prevailing during the past 20 years may entail an increased risk of contracting cancer.

Factors modifying the metabolism and toxicity of cadmium in man must be studied in order to decide whether the increased risk of cadmium toxicity justifies the proposal of more restrictive health-based occupational exposure limits for some population groups.

2.4 Recommendations

The use of both long-term and short-term exposure limits is recommended for the prevention of acute and chronic pulmonary effects. Renal effects can be prevented by controlling the concentrations of cadmium dust and cadmium oxide fumes in air and by periodically ascertaining that blood and urinary levels of cadmium in workers do not exceed the critical values.

Recommended health-based occupational exposure limit for short-term exposure. A short-term exposure level for cadmium oxide fumes and respirable dust of 250 μg Cd/m^3 is recommended for the prevention of acute lung reaction provided the time-weighted average is respected.

Health-based occupational exposure limit for long-term exposure. To prevent adverse pulmonary and renal effects in long-term occupational exposure to cadmium the time-weighted average concentration of airborne cadmium fumes or respirable dust should be below 20 μg/m^3. In the present state of knowledge, a tentative value of 10 μg/m^3 is recommended.

However, to ensure that excessive accumulation in the kidney does not occur, cadmium levels in blood and urine must be considered simultaneously.

Recommended health-based biological limits for cadmium in blood and urine. Normally, the Cd–B level is mainly an indicator of exposure over the last few months. A value of 10 μg Cd/l of whole blood is proposed as the individual critical level if the exposure is regular and long-term.

The Cd–U level can be used to estimate the body burden of cadmium. The individual Cd–U concentration should not be allowed to reach 10 μg Cd/g creatinine, since above this concentration there is some risk of renal dysfunction.

Since the Cd–B and Cd–U levels mentioned above are levels at which renal effects have been observed, and in view of the long biological half-life of cadmium, it is recommended that control measures be applied as soon as the individual concentrations of Cd–U and/or Cd–B exceed 5 μg Cd/g creatinine and 5 μg Cd/l of whole blood, respectively. Consequently these levels should be regarded as health-based biological limits. In view of the current uncertainty regarding the carcinogenic activity of cadmium in man, a revision of the various levels proposed in this report may be required when the results of current epidemiological studies on the carcinogenic properties of cadmium become available.

REFERENCES

1. FLANAGAN, P. R. ET AL. *Gastroenterology*, **74**: 841 (1978).
2. *Chemicals and industrial processes associated with cancer in humans.* Lyon, International Agency for Research on Cancer, 1979 (IARC Monographs on the Evaluation of the Carcinogenic Risk of Chemicals to Man, Vol. 1–20, Suppl. 1), p. 27.
3. LEMEN, R. A. ET AL. *Annals of the New York Academy of Sciences*, **271**: 272 (1976).
4. COMMISSION OF THE EUROPEAN COMMUNITIES. *Criteria for cadmium*, Oxford, Pergamon, 1978.
5. *Environmental health criteria: Cadmium, a summary.* Unpublished WHO document No. EHE/EHC 77.1, 1977.
6. ROELS, H. ET AL. *Lancet*, **1**: 221 (1979).
7. BERNARD, A. ET AL. *European journal of clinical investigation*, **9**: 11 (1979).
8. FRIBERG, L. ET AL. *Cadmium in the environment*, Cleveland, Chemical Rubber Company, 1974.

3. INORGANIC LEAD

This section deals with metallic lead, lead oxides, and lead salts. However, lead salts, such as lead arsenate and lead chromate, in which the anion may possess carcinogenic properties, are not considered from the point of view of such effects. Alkyl lead compounds have also been excluded.

3.1 Summary of metabolism and toxicity

3.1.1 *Metabolism*

Absorption, distribution, retention, and excretion. Inorganic lead compounds are absorbed in the organism via the lungs and gastro-intestinal tract. However, lead salts of organic acids (e.g., lead napthenate and lead stearate) are, to some extent also, absorbed through the skin. Absorption through the lungs is the most important route in occupational exposure while absorption through the gastrointestinal tract is more common in non-occupational exposure.

The details of the mechanism of absorption are not yet fully understood, but the present knowledge regarding lead deposition in the lungs and absorption through the gastrointestinal tract and lungs can be summarized as follows (*1, 2*).

As the particle size decreases, the amount of lead deposited in the lungs also decreases. For example, when the size of the lead particles is 1 μm, about 60% of the inhaled lead dust is deposited in the lungs while when the particle size is only 0.1 μm only 40% of the inhaled dust is deposited. It should be noted that these amounts have been calculted at the normal rate of breathing at rest. Furthermore, increased ventilation decreases the proportion deposited, but the total amount of lead entering the lungs increases.

On an average about $30 \pm 10\%$ of the lead inhaled is absorbed through the lungs. The larger particles are deposited in the upper respiratory tract from where they are transported by the mucociliary escalator to the nasopharynx and swallowed.

The degree of absorption from the respiratory tract is influenced by the differences in solubility of different compounds, the shape, size, in the workroom air, of the particles, and other factors such as smoking habits or the presence of a chronic nonspecific respiratory disease.

It is estimated that about 5–10% of the lead ingested is absorbed via the gastrointestinal tract. However, it is possible that this range may be wider considering the fact that figures of 1.3% to 16% have been published. It is also known that dietary factors, such as low calcium, iron, and protein content of the food, increase gastrointestinal absorption. In infants the gastrointestinal absorption may be greater than in adults.

The absorbed lead enters the bloodstream from where it is distributed to the organs and systems. Redistribution then occurs in relation to the relative affinity of each tissue for lead.

About 95% of the lead in blood is bound to the erythrocytes. The lead present in the organism can be divided into two types—exchangeable fraction and stable fraction. The former consists mainly of lead in blood, in soft tissues, and to a lesser extent, bones. The concentration of lead in blood is in equilibrium with that of lead in soft tissues.

About 90% of the total body burden of lead is present in the bones and teeth, as the stable fraction, which is not accurately indicated by the blood lead level. This form of lead is the result of long-term absorption.

Lead passes through the placenta easily, and fetal blood has almost the same lead concentration as maternal blood (*79*). Lead also passes the blood-brain barrier, although the brain does not accumulate lead.

The elimination of lead takes place mainly through the urine (75–80%) and, to a lesser extent (probably about 15%), by gastrointestinal secretion. Other routes (hair, nails, sweat) account for less than 8% (*2*). Maternal milk contains small amounts of lead (a few micrograms per litre).

The biological half-life of lead is extremely difficult to estimate. The constantly decreasing availability of the major stores of lead in osseous tissue make it virtually impossible to describe the rate of loss from the body in simple terms, but there is no doubt that clearance of half the body burden of lead would require a number of years (*2*).

Concentration of lead in biological media as an indicator of exposure and body burden. Lead can be measured in whole blood, plasma, saliva, urine, hair, deciduous teeth, and bone biopsies. It is generally agreed that the level of lead in blood (Pb–B)—a measure of absorbed lead—is the best indicator of current exposure (*2*). However, in interpreting results one must consider the fact that the blood lead level reflects a dynamic equilibrium between exposure (absorption), retention, release, and elimination. If the body burden is large, its relative impact on the Pb–B is great, and vice versa. Under steady state conditions, e.g., those prevailing in the general population or during long-term, unchanged occupational exposure, the blood level gives a good picture of current exposure, but shortly after changes in exposure intensity the Pb–B becomes a poorer indicator. For example, after the start of occupational exposure, it takes about two months for the Pb–B level to reach a steady state, and the level falls slowly after the termination of exposure (*2*); the half-life of lead in blood is about 2–4 weeks.

It is generally agreed that the concentrations of lead in air, food, and water are less relevant for assessing health hazards than the amount actually absorbed, and this is what is actually reflected by the Pb–B level. For this reason the Pb–B level, rather than the concentration of lead in workroom air, will be used as the primary measure of exposure in this context. Measurements of lead in air are, however, of primary importance in controlling exposure in the workplace.

Past exposure, or body burden, can be estimated by measuring the concentrations of lead in deciduous teeth (in children) or by measuring the amount excreted in urine after provocation with a chelating agent (*3*). Measurement of lead in hair allows the estimation of exposure over a period of one to several months, but this method may not be reliable because external contamination usually interferes and the results may not be accurate, especially in occupational exposure. The excretion of lead in urine after chelation reflects the concentration of the metal in the soft tissues. However, none of these methods is useful in measuring past *occupational* exposure.

The Pb–B level does not accurately indicate the body burden, although some relationship between the two has been observed; rarely, if ever, will the level decrease to completely normal levels in persons whose body burden is large.

Normal levels of lead in the blood. Many studies indicate that the average Pb–B level in adults without occupational exposure is usually in the range of 100–250 μg/l (*2*). Scandinavian populations have exceptionally low levels. A mean value of 85 μg/l was found in 50 women from southern Sweden (*4*), while the mean values of various Finnish population groups ranged from 79 μg/l (rural females) to 114 μg/l (urban males) (*5*). A low mean concentration

(99 μg/l) has also been found, among 900 Iowa, USA, women, with a range of 20–290 μg/l (*6*). The other extreme occurs in northern Italy and in France, where high mean values above 200 μg/l and even 300 μg/l have consistently been measured. Significant variations exist among other countries as well (*2*).

The Pb–B level does not increase with age in adults. As a rule, males have slightly higher Pb–B levels than females. Similarly, urban populations, especially those exposed to traffic exhaust fumes, have higher Pb–B levels than do rural populations. This rule applies both in countries with high and in those with low basic levels (*5*). Apart from exhaust fumes from petrol engines, contaminated drinking-water, wine, and food (e.g., lead dissolved from ceramic glazes) are the most important sources of non-occupational exposure (*7*). Increased lead absorption has also been recorded in people living in the vicinity of lead smelters (*2*).

3.1.2 *Toxic effects*

Haematopoietic system. The haematological effects of lead can be attributed to the combined effect of (1) the inhibition of haemoglobin synthesis, and (2) the shortened life-span of circulating erythrocytes. These effects may result in anaemia.

In man, disturbances in haem synthesis become manifest when abnormal concentrations of haem precursors appear in blood and urine. Lead also interferes with iron metabolism and with globin synthesis in erythrocytes. Of the enzymes involved in haem synthesis, lead inhibits δ-aminolevulinate dehydratase[1] (ALAD) and haem synthetase (ferrochelatase), and probably also coproporphyrinogen decarboxylase, either by direct action or by negative feedback mechanism, and sometimes by both (*2, 3, 7*). This enzymatic inhibition results in increased excretion of δ-aminolevulinic acid (ALA) and coproporphyrin III in urine and accumulation of protoporphyrins in the erythrocytes.

The life-span of circulating erythrocytes also becomes shortened, and together these two defects finally cause mild anaemia and sometimes microcythaemia (*3*). The toxic effect of lead on haematopoiesis manifests itself in morphological changes in normoblasts in

[1] Porphobilinogen synthase (E.C. 4.2.1.24).

the bone marrow. In peripheral blood, reticulocytosis and stippled erythrocytes can also be seen; however, both of these are nonspecific phenomena.

Neurological effects. High lead exposure causes encephalopathy, the classical signs and symptoms of which are ataxia, coma, and convulsions. Many survivors of lead encephalopathy sustain residual brain damage which is manifested by mental and/or neurological impairment. Recent studies indicate that milder exposure, which does not cause the above-mentioned serious disorders, may give rise to subjective symptoms indicative of central nervous effects (*8, 9, 10*), and impaired psychological performance has been seen among adult lead workers (*8, 11, 12, 13*). However, owing to the subtle nature of such effects, they can be shown only on a group basis and a standardized methodology is an absolute requirement.

Psychological disturbances, such as learning difficulties, behavioural changes, and intelligence defects, have also been observed in children who did not have a history of encephalopathy but showed evidence of increased lead absorption (*2, 14*). These studies, which suggest that there are effects on the central nervous system of young children when the Pb–B ranges between 200 and 400 μg/l, will be of relevance when exposure limits for female workers in the fertile age are discussed (see 3.2.7 below).

Peripheral nervous defects range from paresis to slight functional impairment detectable only by sensitive electrophysiological techniques (*2*). Until recently, demyelination of the myelin sheaths was considered to be the main lesion. However, it has been shown (*15*) by electron microscopy that axonal degeneration of the nerves is more prominent. A presynaptic block may also be a factor since reduction of the end-plate potential has been described (*3*).

Renal effects. The effect of long-term exposure to lead on the kidneys consists of a nonspecific nephropathy characterized morphologically by intense interstitial fibrosis, tubular atrophy, and dilatation with relatively late involvement of glomeruli. Tubular hypofunction may also occur, and renal intranuclear inclusion bodies have been demonstrated at exposure levels much lower than those leading to clinical nephropathy. In asymptomatic adults with increased Pb–B levels, increased blood urea nitrogen levels, and disturbances in renal clearance have been reported (*16, 17*). Clinically there may be hyperuricaemia and sometimes gout.

Children who have had intense short-term exposure may develop the Fanconi syndrome (*2, 3*).

Gastrointestinal effects. The best known gastrointestinal effects caused by lead exposure are constipation or diarrhoea, epigastric pain, nausea, indigestion, and loss of appetite; those may culminate in colic (*7*).

Effects on liver. There is no definite evidence showing that the liver is affected by lead, although there have been a few reports indicating that the liver is affected in heavily exposed persons (*2*).

Effects on the cardiovascular system. Some reports suggest that high exposure may exert a toxic action on the heart. Clinical cases of lead poisoning sometimes show symptoms of myocarditis. There is, of course, the possibility that the appearance of these symptoms with lead poisoning is purely coincidental; however, in many cases electrocardiographic symptoms disappeared with chelation therapy, suggesting that lead may have been the etiological factor (*18, 19, 20*). According to a British study, the cerebrovascular mortality rate, when compared with the expected rate, was high among heavily exposed lead workers. These men had been exposed to lead during the first quarter of this century when working conditions were much worse than those prevailing today in most countries; no similar increase in the mortality rate for men employed more recently could be found (*21*).

Effects on reproduction. Numerous reports in the older literature on stillbirths and miscarriages among women working in lead trades suggest that fertility is impaired by high exposure (*2*). The reproductive capability of men may also be impaired by high-level exposure; this is evidenced by an increased frequency of asthenospermia, hypospermia, and teratospermia. However, there are no dose-related data at present on the effects of low-level lead exposure on reproduction. Therefore, these effects cannot yet be considered in the recommendation of health-based limits. Fetotoxic effects may, however, be considered on the basis of current evidence.

Mutagenicity. The literature is controversial with regard to chromosomal abnormalities in subjects exposed to lead (*2*). Several of the studies published have not been able to exclude concurrent exposure to other metals. Two recent studies (*22, 23*), have added more evidence in favour of the hypothesis that lead may induce chromosomal abnormalities.

In the first study (*22*), 20 workers with lead poisoning were examined. The study excluded workers (*a*) who had been exposed to known mutagenic agents, (*b*) who had had X-ray examinations or treatments during the last two years, (*c*) who had had recent viral infections, (*d*) who had used medicinal preparations known to be mutagenic, and (*e*) who had a history of treatment with calcium disodium edetic acid. The Pb–B level at the time of examination was in the range of 500–1000 μg/l. A significant correlation was observed between the number of abnormal metaphases and the current Pb–B levels.

The second study (*23*) established a dose–effect relationship between the Pb–B level (mean of several determinations at different periods of time) and chromosomal aberrations in 26 lead-exposed smelter workers. In this study three different exposure groups were defined; a low exposure group (mean Pb–B 225 μg/l), a medium exposure group (mean Pb–B 390 μg/l) and a high exposure group (mean Pb–B 650 μg/l). The frequency of gaps, chromosomal aberrations, and chromatid aberrations increased as the mean Pb–B level increased. The most sensitive abnormality was the increased frequency of gaps, which was statistically significantly different from the reference group even in the low exposure group. For all three types of aberrations the frequency increased with rise in the Pb–B level, but this increase was statistically significant for chromosomal aberrations only. The merit of this study lies in the longitudinal exposure data, but unfortunately the authors did not discuss the possibility of concurrent exposure to other metals (e.g., arsenic) that is known to occur in the vicinity of the particular type of smelter studied.

Although the cause–effect relationship between Pb–B and chromosomal aberrations cannot be considered completely established yet, the two above-mentioned studies speak strongly in favour of such a relationship. According to the second study, an increased risk of chromosomal damage may be found even at mean Pb–B levels of about 250 μg/l. However, the possibility of confounding exposure to other metals renders the interpretation of this effect threshold uncertain; on the other hand, the possibility of a confounding exposure causing effects similar to those shown by lead is remote since the workers examined were particularly occupied with lead smelting.

Finally, it should be added that, using an *in vitro* DNA synthesizing system, Sirover & Loeb (*24*) could show that lead chloride caused a significant decrease in the fidelity of DNA synthesis.

Carcinogenicity. Animal studies have shown that a number of lead compounds induce benign and malignant tumours in several species (*2*). However, no conclusive epidemiological study has yet been published in favour of lead being a human carcinogen (*2, 25*). The only epidemiological hint of cancer due to lead exposure comes from a survey made by Cooper (*26*). In this study an excess of lung cancer was found in heavily exposed secondary lead smelters (standard mortality ratio 121) and battery plant workers (standard mortality ratio 128). The standard mortality ratio for all cancers was also elevated among the battery plant workers (standard mortality ratio 136), but not among the smelters (standard mortality ratio 89). However, concurrent exposure to other metals may be partly responsible for these results. No dose-response pattern emerged; therefore the results are not conclusive.

Other effects. According to a USSR study (*27*), changes in blood electrolytes occurred in 153 workers of a radioassembly plant, exposed to lead at concentrations around 10 $\mu g/m^3$. For example, the quotient of sodium/potassium concentration in the erythrocytes was 0.25 ± 0.006 as opposed to 0.22 ± 0.008 in a control group ($P < 0.02$) and the calcium/phosphorus quotient was 3.39 ± 0.1 as against 2.51 ± 0.17 in plasma and 1.16 ± 0.06 as against 0.78 ± 0.06 ($P < 0.001$) in the erythrocytes, respectively.

Effects relevant to health-based occupational exposure limits. In making recommendations on health-based occupational exposure limits, effects that can be considered as relevant should (1) represent functional changes that have a potential to cause an impact on health and (2) represent early indications of exposure. Effects that occur at high exposure levels will be automatically prevented if early indicators of exposure are used as guidelines. This explains why, for example, nephropathy has not been considered in this context. On the other hand, some effects measurable by, for instance, sensitive biochemical tests, may occur at exposure levels that are lower than the ones at which health impairment occurs. If these effects do not cause functional impairment, then they can be judged to be without any health significance and need not be considered in this context.

A Subcommittee on the Toxicology of Metals convened by the Permanent Commission and International Association on Occupational Health in 1974 considered the interference of lead with haem synthesis to be the earliest relevant adverse effect (*3*). The Subcommittee felt that further research might produce methods for measuring neurophysiological and/or neurochemical disturbances that were believed to occur before the effects on haem synthesis. However, since the meeting of the Subcommittee, several studies have been published showing that the effects on the nervous system occur when the levels of lead in the blood are of the same order of magnitude as those that cause increases in levels of haem synthesis intermediates in blood and urine. Furthermore, according to a recent study (see *Mutagenicity* above) chromosomal aberrations may also occur at similar levels. Although there are still some doubts about their conclusiveness, these effects may have to be included in relevant effects once they are confirmed by further research.

3.2 Relationship between exposure and relevant health effects

3.2.1 *Relationship between biological and environmental exposure indicators*

There is little precise information on the relationship between the concentrations of lead in air and in the blood in subjects who are occupationally exposed. Since the level of Pb–B is primarily a product of lead in air some relationship between the two must exist in the sense that low levels of lead in air give rise to low Pb–B levels and high air levels cause the lead concentration in the blood to reach correspondingly high levels. However, as discussed by the Occupational Safety and Health Administration in the USA (*28*), there are several reasons why the relationship between lead levels in air and lead levels in the blood cannot be expected to be very close under ordinary working conditions. First, the Pb–B level in non-occupational exposure varies among individuals and this causes scatter around the curvilinear air-lead and blood-lead regression line; it also alters both the slope of the curve and its intercept of the *y*-axis. This results in overestimates of Pb–B levels at low concentrations of lead in air and underestimates at high concentrations. Secondly, in view of the differences in particle size (resulting in different respiratory volumes) between workplaces, it is difficult or impossible to produce universally applicable regression curves.

Thirdly, in order to be representative, the measurements of concentrations of lead in air should be taken over a long period of time and the sampling site should correspond to the breathing zone. Failure to meet these criteria results in increased scatter and sometimes in systematical errors. Finally, analytical errors may cause both scatter and bias.

Some of these sources of error can be minimized by means of careful planning of studies, but other sources may be difficult to avoid (e.g., it may be impossible to construct "universal" regression curves for all types of work). Unfortunately, the studies published thus far have not succeeded in overcoming the difficulties described above. In general, the following flaws are inherent in them.

(1) The particle size distribution and the chemical form of the lead in the air are not considered. Usually too little attention is paid to the fact that non-respirable particles are swallowed and partly absorbed via the gastrointestinal tract. If only respirable airborne lead is considered, the exclusion of non-respirable particles causes underestimates of the total amount absorbed, which vary from one exposure situation to another.

(2) The measurements of airborne lead concentrations are usually of short duration in relation to the true exposure time. Owing to wide fluctuations in dust concentrations over periods of days, weeks, and months, short-time (even 8 h) samples are usually not representative.

(3) The individual blood lead level of the workers prior to occupational exposure is not usually known, and even if an allowance is made for the non-occupational blood lead level the scatter is not eliminated.

(4) Differences in physical activity, resulting in corresponding differences in the breathing volume, are not usually considered. The fact that some workers wear respirators while working may also influence the relationship between blood lead and concentration of lead in air, as found during a short period of time.

(5) The contribution of exposure from contaminated hands, cigarettes, food items, etc., is not usually accounted for. According to a recent study, lead workers who smoke have in general higher blood lead levels than non-smokers; this difference was interpreted as resulting from contaminated cigarettes (*29*).

The study of Williams et al. (*30*) on the relation between blood lead and lead concentration in air is the one that has received most attention. In the WHO Environmental Health Criteria publication on lead (*2*), these data have been used to estimate the potential contribution of lead in air to Pb–B in subjects who were occupationally exposed. However, in 1978, Williams (*31*) pointed out that a 23% systematic error was discovered in the calibration of the personal sampler. Furthermore, each man was studied for two weeks only, and there was no evidence showing that the levels of lead in air were typical of his normal exposure. The particle size was not measured and the share of non-occupational exposure was unknown. For these reasons the calculations made in the Environmental Health Criteria on lead, showing an average of 10 μg/l in Pb–B per 1 $\mu g/m^3$ increase in the concentration of lead in air may not be correct, especially at high levels of lead in air. In fact, a power function curve constructed to fit the plot of Pb–B against the time-weighted average concentration of lead in air showed that the former increases curvilinearly with the latter at high levels of lead in air; i.e., the expected increase in Pb–B per unit increase in the concentration of lead in air shrinks continuously as the Pb–B levels increase.

Moreover, if Williams's data should constitute the actual relationship, then these data are only indicative of the special work situation and such a relationship cannot be considered as representative of occupational exposure to lead in general.

Recent data from a survey of storage battery plant workers in Muncie, IN, USA, involving about 5000 estimations of Pb–B, showed a weak relationship ($P < 0.05$) between lead in air as measured by personal samplers, and the levels of Pb–B (*32*). The most relevant data are shown in Table 3.

As can be seen from the table, only 6% of the Pb–B values exceeded 600 μg/l even when the lead concentrations in air were above 150 $\mu g/m^3$. Ninety-five percent of the values were below 600 μg/l when the air lead concentrations were below 100 $\mu g/m^3$. However, a close relationship could not be established between the two. For example, at air lead concentrations in excess of 150 $\mu g/m^3$ only 23% of the Pb–B levels exceed 500 μg/l, whereas a 10% increase was observed at concentrations below 50 $\mu g/m^3$. The most prominent difference between the air lead concentrations of < 100 and < 200 $\mu g/m^3$, respectively, was that the proportion of lead workers with Pb–B levels below 400 μg/l was 37% at the higher concentration and 51% at the lower. However, this study also suffers from the errors

Table 3. Relation between the concentration values of lead in air and levels of lead in blood in storage battery plant workers

Levels of lead in air (μg/m³)	Levels of lead in blood							
	<400 μg/l		<500 μg/l		<600 μg/l		≥600 μg/l	
	No. of samples	%	No. of samples	%	No. of samples	%	No. of samples	%
<200	1 721	37	3 482	74	4 403	94	290	6
<150	1 328	41	2 509	77	3 054	94	201	6
<100	761	51	1 253	84	1 432	95	63	5
<50	315	64	449	90	490	99	6	1

discussed earlier in this section; this fact renders the interpretation of the data difficult. In particular, the fact that single measurements of air lead concentrations were paired with Pb–B measurements made within less than a month of air sampling is a serious flaw.

A few laboratory studies have also been performed. Although they were conducted under controlled laboratory conditions, the results are difficult to relate to occupational situations. Coulston et al. (*33*) exposed 14 male volunteers to a lead oxide aerosol for 23 hours per day at an average concentration of 11 μg/m³ for 7 weeks. They found that the contribution of air lead to the average Pb–B levels was approximately 14 μg of lead/l of blood per 1 μg/m³ of air. In another study male volunteers were exposed to an air lead concentration of 3.2 μg/m³ (*34*). Their average Pb–B level increased from 180 μg/l to 250 μg/l. In other words, 1 μg of lead per m³ caused an increase in the average Pb–B level of approximately 5 μg/l when adjusted to an equivalent 40-hour-per-week occupational exposure (*34*). In other experiments with three volunteers, Rabinowitz et al. (*35, 36, 37*) found that 1 μg of lead per m³ of air increased the Pb–B concentration by approximately 10 μg/l (based on a 40-hour-per-week equivalent occupational exposure). Although a relationship seems to exist between lead in air and lead in the blood, the above-mentioned experiments cannot be considered conclusive because Rabinowitz et al. studied only three individuals and Coulston et al. (*33, 34*) used a lead aerosol which was not representative of the variety in particle size distribution in general or in the workroom air. In measuring occupational exposure one must remember, as mentioned above, that the lead particles occurring in air vary from one type of work to another. Hence it can be concluded that neither the fields surveys nor the laboratory studies published so far provide reliable data for the construction of a regression curve between lead in air and Pb–B.

However, the US Occupational Safety and Health Administration (*28*) made some calculations based on the Bernard model, as applied by the Center of Policy Alternatives (Massachusetts Institute of Technology). These calculations suggest that at an air lead concentration of 50 μg/m^3, 0.5% of workers will have Pb–B levels exceeding 600 μg/kg of body weight, 5.5% will have levels between 500 and 600 μg/kg, and 23.3% will have levels above 400 μg/kg. Hence in all, 29.3% of the workers will have Pb–B levels above 400 μg/kg at any one time when there is a uniform exposure level of 50 μg/m^3.

The Bernard model referred to above considered the variability around the mean blood lead concentration in determining the distribution of blood lead levels at a given air lead exposure. This model predicts that at a given concentration of lead in air (50 μg/m^3) the mean Pb–B level in a population will be approximately 350 μg/l. Furthermore, if one assumes that the mean Pb–B level of a population is 150 μg/l before entering the workforce, then the average increase in Pb–B for every 1 μg/m^3 increase in the concentration of lead in air would be about 4 μg/l, calculated on the basis of 40 h per week exposure to air lead concentrations of up to 50 μg/m^3. It should be noted that this is in close agreement with the independent estimate by Bridbord (*38*). His study, like that of Coulston et al. (*34*), indicates that for every 1 μg/m^3 increase in air lead the Pb–B rises by about 5 μg/l in a 40-hour-per-week average exposure to air lead concentrations of 0–50 μg/m^3. These predictions are also in agreement with the data of Noweir et al. (*39*), who demonstrated an average Pb–B level of 348 μg/l in a group of workers exposed for 60 h to an average air lead concentration of 30 μg/m^3 (45 μg/m^3 based on a 40-h working week), compared to a control group assumed to have had no occupational exposure; the blood lead concentration of the control group was 190 μg/l.

The method of deriving an air lead standard indirectly from data based on the relationship between Pb–B and symptoms and signs of toxicity may be open to criticism. There are some studies directly relating air lead levels to effects, but considering the above discussion close relations cannot be expected.

Most, if not all, dose–effect and dose–response data on lead rely on Pb–B levels, and because the correlation between the two is so weak, the Pb–B level must be used as a starting point for the definition of a health criterion. Moreover, in the case of lead exposure, the Pb–B level describes total (occupational and non-occupational) uptake, and this total exposure is decisive in estimating the health risk.

Once the health-based biological limit has been defined for Pb–B, a corresponding air lead concentration value can be estimated. This value would prevent the great majority of the workers from exceeding the recommended Pb–B limit. Owing to the weak relationship between the two, the confidence intervals of the regression are wide.

3.2.2 *Analytical problems of blood lead measurement and their effect on interpretation of results*

It is well known that Pb–B determinations are highly vulnerable to methodological errors. Especially in the past, many laboratories performed the measurements unsatisfactorily. The errors of even the most experienced laboratories can reach as much as 10% and those of less experienced laboratories are much higher; for example, an inter-laboratory comparison between 66 European laboratories employing different analytical techniques showed that the results differed by a factor of up to 10 (*40*). Similar examples have been reported from the USA (*41*). However, it seems that these regrettable results have had good effects since they have prompted intensified interlaboratory method control programmes and other measures to increase the precision and accuracy of the analytical methods. There are many regular interlaboratory checking programmes in operation now around the world. For example, the Nordic countries have had an intercomparison programme since 1971, and several other countries, such as the member countries of the Commission of the European Communities have organized similar control programmes. A biological monitoring programme has also recently been initiated by UNEP and WHO. The need for strict quality control has been emphasized repeatedly. In general the results of the intercomparison programmes indicate a considerable improvement in analytical results. It is believed that the inaccuracies depend more on human factors than on the method employed.

Several methods now exist that have proved to be valid and reliable in experienced hands. Of them the traditional dithizone (diphenylthiocarbazone) method, atomic absorption spectrometry, and anodic stripping voltametry as well as some new modifications of polarography can be mentioned, but other good techniques also exist. Avoidance of contamination, careful handling of the samples, and frequent intra- and interlaboratory checks are more important for ensuring accuracy than the method as such.

In spite of the improvements that have been made in analytical methods, only the data from a few recent studies can be fully accepted for relating Pb–B levels to toxic effects. However, this still does not usually permit the comparison of results from different laboratories because many discrepancies with regard to dose–effect relationships arise from comparison of incompatible results. The following considerations rely mainly on studies from laboratories with well established accuracy and especially on studies showing quality control results. However, the problem referred to above can never be completely eliminated.

3.2.3 *Exposure-effect relationships*

Haematological effects. Accumulation of protoporphyrin IX in erythrocytes is the earliest of the haematological effects. This effect occurs as a result of inhibition of ferrochelatase (haem synthetase) activity in the bone marrow. Some of the analytical methods in use measure the protoporphyrin IX (PP) concentrations, while others assess total free erythrocyte porphyrins (FEP), which according to recent studies, in fact, are not "free" but exist as zinc protoporphyrin IX (ZnPP). This complex can be readily detected as such in dilute whole blood by fluorometry (*2, 42, 43*).

Several studies have shown that a close relationship exists between Pb–B and the erythrocyte porphyrin content (*44–54*). Although different methods yield different results it is not important in principle whether FEP, ZnPP or PP is measured, since about 90% of the excess of porphyrins due to lead exposure are composed of protoporphyrin IX (*55*).

According to Piomelli et al. (*45*) the mean "normal" FEP level is 469 ± 149 µg/l erythrocytes with a range of 220–870 µg/l. Other studies using the same method have found approximately similar values, e.g., Hotz et al. (*56*) calculated values of 570 ± 120 µg/l erythrocytes for young girls whose mean Pb–B level was 126 ± 30 µg/l and Tomokuni & Ogata (*57*) found values of 430 ± 150 µg/l erythrocytes for male workers whose Pb–B level was 100 ± 20 µg/l. However, methodological differences render a comparison of absolute PP (or FEP) values difficult. In addition, the average Pb–B level of the "normal" population may also determine the "normal" FEP level, since there is an interrelationship between the two.

As the Pb–B levels increase, the PP concentration in the erythrocytes also begins to increase. However, close correlations are found only under steady conditions of exposure. Since the normal average life-span of an erythrocyte is about 120–130 days, their entire replacement in the blood requires that length of time. This explains why there is a time-lag between the increase in erythrocyte PPs and the rise in the Pb–B level; this weakens the correlation between the two during the first four months of increased exposure (*2*). The PP content of the erythrocytes also remains increased out of proportion to current Pb–B levels after the termination of occupational exposure (*49*). The following considerations therefore concern steady exposure conditions only.

According to the consensus of three scientific studies (*2, 3, 58*), the PP starts to increase at Pb–B levels of about 300–400 μg/l in men and at about 250–300 μg/l in women. Recent data indicate that young children react at even lower Pb–B levels (*59*). The slope is steeper for females than for males (*2, 60*).

Fig. 1 and 2 show the relationship between the erythrocyte ZnPP content and the Pb–B level and the EPP (erythrocyte PP) content and the Pb–B level of Italian male workers respectively (*51*). There are some differences between the ZnPP and the EPP values, but the most

Fig. 1. Relationship between erythrocyte zinc protoporphyrin (ZnPP) content and Pb–B in 191 male lead workers (*51*)

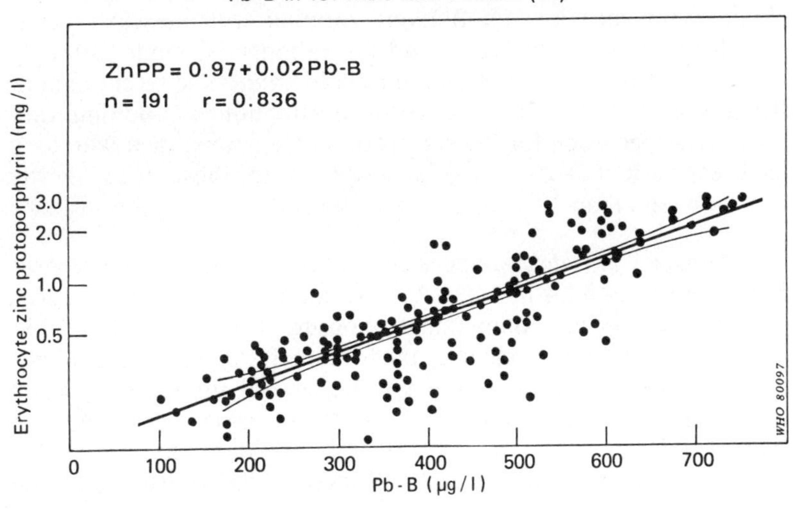

Fig. 2. Relationship between erythrocyte protoporphyrin (EPP) content and Pb–B in 191 male lead workers (*51*)

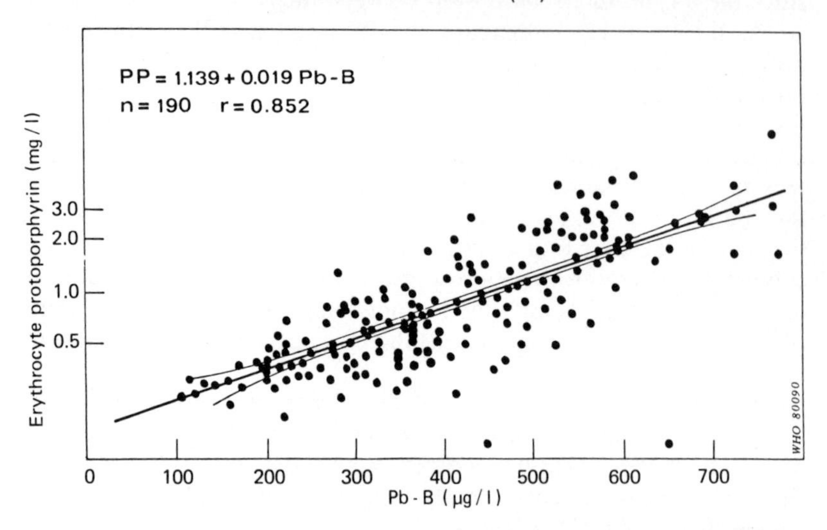

interesting feature is the absence of any effect threshold, irrespective of the protoporphyrin method used. According to Alessio (private communication), the linear correlation coefficient between the ZnPP and Pb–B levels below 350 µg/l was 0.50 ($P < 0.001$) and the exponential coefficient 0.55 ($P < 0.001$). For EPP the coefficients were 0.52 and 0.59, respectively. The fact that significant correlations were obtained for such low Pb–B levels, together with the close correspondence between the linear and the exponential model, strongly suggests that there may not be a threshold value. The recent data of Roels et al. (*60*) seem to support this assumption; in addition, they show a steeper slope for the regression curve among men with long lead exposure (10–20 years) as opposed to those with shorter exposure (less than 10 years).

Changes in delta-aminolevulinic acid (ALA) metabolism. Measurement of ALA in urine (ALA-U) has been used as a biological test for monitoring occupational exposure to lead (*2, 7, 23, 54, 61, 62*). It is useful in evaluating the health effects but it is considered to be an indirect indicator of exposure. Several studies have shown that increased concentration of ALA in urine correlates with increased Pb–B levels; however, the threshold of the mean ALA increase lies between the mean Pb–B levels of 350–450 µg/l (*2, 3, 60,*

62). Furthermore, Nordberg has shown that even at Pb–B levels above 400 μg/l the increase in ALA excretion is not very significant; the correlation coefficients are in the range 0.5–0.7 (3). According to Roels et al. (48, 60) females seem to be more sensitive to this effect than males. In the first study (48) a correlation between Pb–B and ALA-U occurred for females even in the Pb–B range of 200–500 μg/l, and the second investigation (60) showed that the approximate mean group Pb–B levels at which significant increases in ALA-U started to occur were 350 μg/l for females and 450 μg/l for males.

Changes in coproporphyrin metabolism. The pattern of coproporphyrin excretion in the urine follows closely that of ALA although the latter is more specific for lead (3, 62). The no-effect level is around 350–400 μg Pb/l of blood. Because of the similarities between the effects of Pb upon the excretions of ALA and coproporphyrin, only the former will be discussed further.

Other haematological effects. These include inhibition of δ-aminolevulinate dehydratase (ALAD) activity, shortening of the erythrocyte life-span, decrease in the haemoglobin concentration, depression of erythrocyte Na^+ and K^+-ATPase activity, and decrease in the reduced glutathione content of blood (2, 7, 63). Since all these effects, with the exception of ALAD activity inhibition, occur at Pb–B levels similar to or higher than those which induce increases in the erythrocyte PP concentration and the excretion of ALA and coproporphyrin in urine, the dose–effect relationship data for them are not relevant in this context.

In one study (64), slight haematological effects were found among 158 females exposed to lead in air concentrations of 7–13 μg/m³ for more than 10 years. The mean ALA excretion was 2.7 ± 0.5 mg/g creatinine as compared to 0.27 mg/g creatinine in a control group. Also the haemoglobin level and the erythrocyte count were lower than those in the control group. However, the fact that these differences were found in a group which had had more than 10 years' exposure suggests that the body burden rather than current exposure levels had caused the effects in question.

Neurological effects. Relating neurological effects to Pb–B levels is more complex than relating haematological effects to P–B levels because in the case of neurological effects there is a time factor, i.e., the Pb–B level measured in connection with the examination of the nervous system may not be the one that actually caused the detected neurological disturbance. This lack of representativeness is especially marked for behavioural disorders in adolescents whose relevant exposure may have taken place several years ago. Moreover, peripheral nervous damage may also persist for a long time. In such situations some measure of the body burden that relates past exposure would certainly be a better indicator than the current Pb–B level. To assess the body burden, some researchers have measured the lead content of hair and deciduous teeth or the urinary excretion of lead after chelation. While these methods are important in making cause-effect inferences, it is difficult to relate them to exposure intensities in terms of the Pb–B level.

In making recommendations on exposure limits, it may therefore be wise, even for neurological effects, to emphasize studies where the Pb–B level has been used as the indicator of exposure, but reservations must be made for the time factor. Close relations between current Pb–B levels and neurological effects can be expected to occur only when no major changes in the exposure intensity have taken place over time. Ideally past records of Pb–B levels should be used, if available, for calculating average exposure intensity and for recording peak exposures. In the actual working environment, long-term fairly stable exposure is not uncommon but unfortunately few studies have taken advantage of this situation in assessing neurological effects.

Electrophysiologically detectable abnormalities in nervous function may occur in the absence of clinical neurological signs (*2*). The major neurophysiological disturbances consist of slowing of the motor conduction velocity (especially of the slower fibres), slowing of the sensory conduction velocity, and electromyographic abnormalities (such as fibrillations and a diminished number of motor units in maximal contraction).

In a study of storage battery workers whose Pb–B levels had not exceeded 700 μg/l, Seppäläinen et al. (*65*) found approximately 5–10% slowing of the maximum motor conduction velocities of the median and ulnar nerves. The conduction velocity of the slower fibres of the ulnar nerve, and the sensory conduction velocity (SCV) of the median nerve also showed a similar reduction in comparison to a control group. At the examination the mean Pb–B level was 400 μg/l with a standard deviation of ± 99 μg/l. Fig. 3 shows a comparison between the conduction velocities of the slower nerve fibres of the ulnar nerve of patients with clinical lead poisoning, lead workers whose Pb–B levels had not exceeded 700 μg/l, and two control groups (*65*). As can be seen, the lowest average conduction velocity of the slower nerve fibres was found in the poisoned group, while those with a maximum Pb–B level below 700 µg/l came next. Both control groups had significantly higher velocities. An individual dose–effect relationship for the conduction velocities of the slow nerve fibres was not quite evident, but on a group basis such a

Fig. 3. Conduction velocities of the slow motor fibres of the ulnar nerve of lead-poisoned subjects (Pb–B not exceeding 700 µg/l), lead-exposed workers, and a group of controls (*65*)

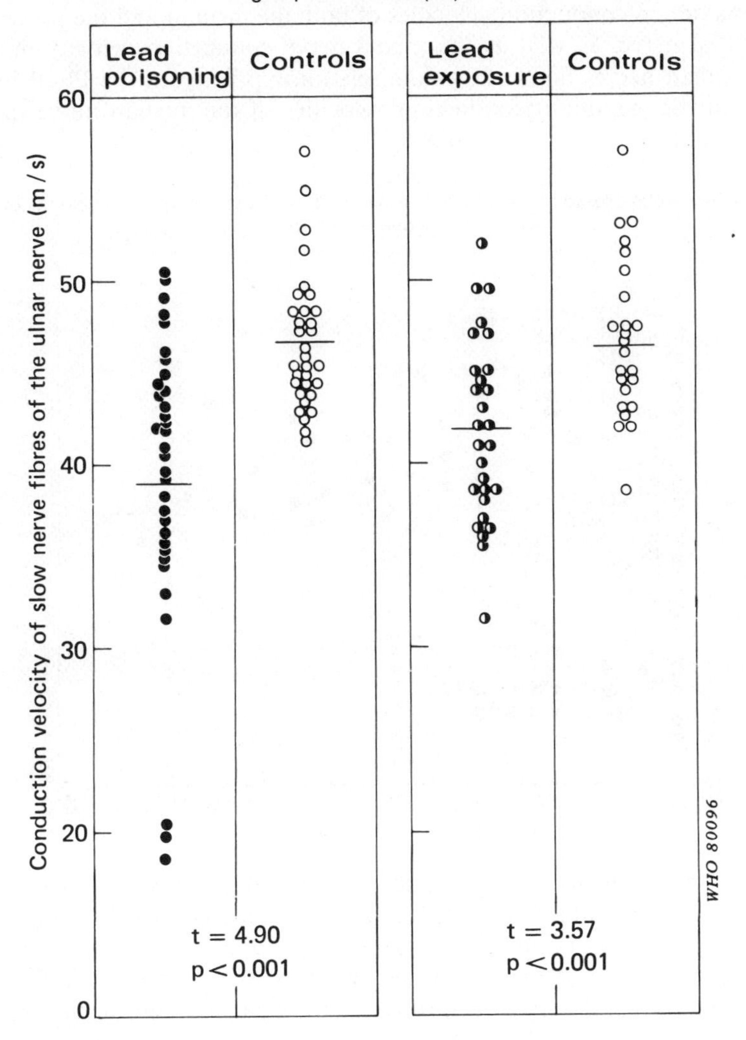

relationship seemed to emerge (Fig. 3). The average conduction velocity of the slow nerve fibres reduced by about 5–10%, but in asymptomatic workers 12 out of 26 had conduction velocities that were below 42 m/s, i.e., the lower limit for 53 out of 54 controls. In some lead-exposed workers the conduction velocities were up to about 20% below this lower limit.

In 1976, Araki & Honma (*67*) found a statistically significant dose–effect relationship between the current Pb–B levels and the maximum conduction velocities of both the median and the posterior tibial nerve, as well as the mixed nerve conduction velocity of the median nerve. Fig. 4 shows the relationship between the Pb–B level and the maximum conduction velocities of the median nerve in 38 subjects.

Fig. 4. Relationship between Pb–B and maximum motor conduction velocity of the median nerve (*67*)

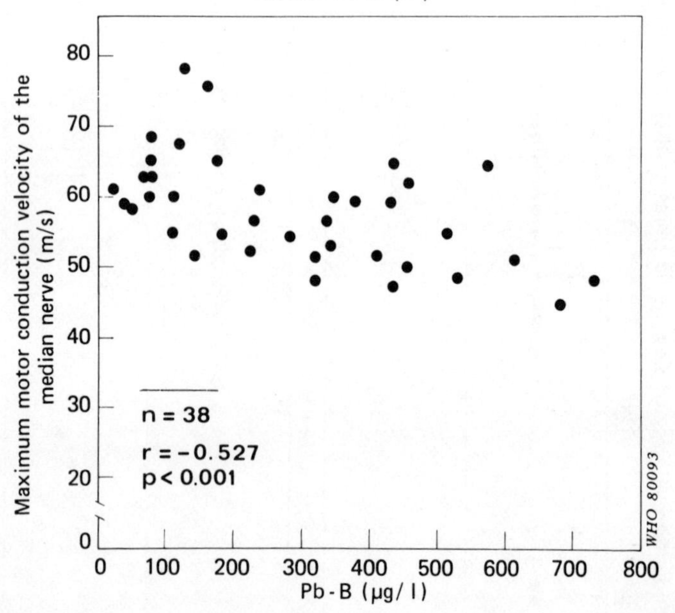

In recent study, Seppäläinen et al. (*68*) confirmed that lead workers do indeed have reduced nerve conduction velocities. This study shows dose–effect relationships for several conduction velocities in a group of 78 lead workers. Most workers had been regularly monitored during their entire exposure period. Seventeen workers had had Pb–B levels exceeding 700 μg/l, while the others had Pb–B levels below that level. A group of 34 workers without any present or past occupational exposure to lead was used as control. During the study statistically significant correlations were found between the maximum Pb–B levels and the nerve conduction velocities of the median nerve, i.e., the maximum conduction velocity, the sensory conduction velocity, the distal sensory conduction velocity, and the motor distal latency. In addition, the conduction velocities of the slower nerve fibres, the sensory conduction velocity of the ulnar nerve, and the maximum conduction velocity of the posterior tibial nerve were also correlated with the maximum Pb–B level and the sensory conduction velocity of the median nerve (Fig. 5). The current and the time-weighted average Pb–B levels were also correlated with measures of nervous function; however, this was to be expected since there were high intercorrelations between the different Pb–B indicators used in this study.

An attempt was also made to define the no-effect level of impairment of nervous function by dividing the lead-exposed group

Fig. 5. Relationship between maximum Pb–B and sensory conduction velocities of the median nerve (*68*)

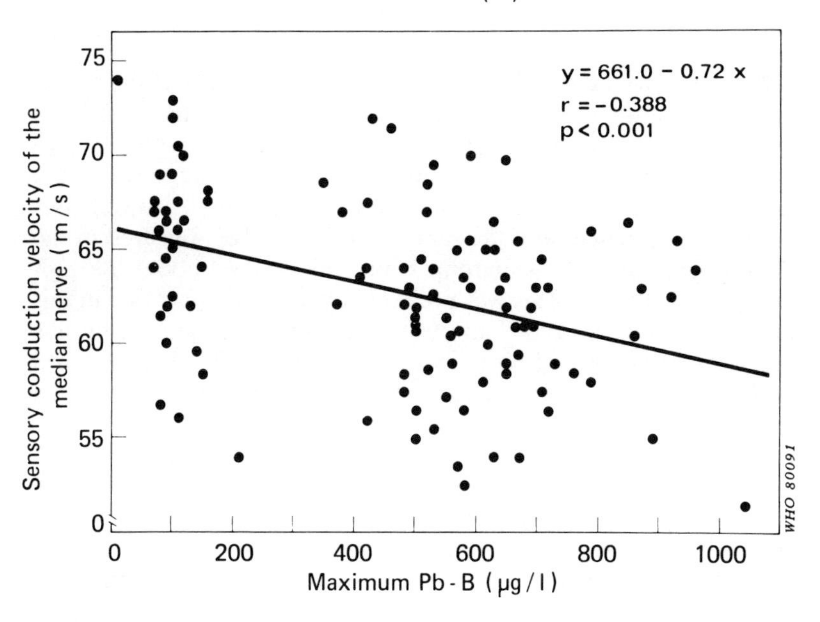

into subgroups on the basis of maximum Pb–B levels. The differences of the means were tested by the *t*-test. Table 4 shows the results.

Table 4. Probability values of differences in the nerve conduction velocities of lead workers in various exposure categories (on the basis of maximum blood lead levels) and a control group comprising 34 non-exposed subjects. Only velocities with at least one significant difference are shown (*68*)

Nerve and conduction velocity	Pb–B			
	400–499 μg/l 11 workers	500–599 μg/l 28 workers	600–699 μg/l 19 workers	≥700 μg/l 17 workers
Median nerve				
Maximum conduction velocity	NS	NS	NS	<0.02
Sensory conduction velocity	NS	<0.005	<0.01	<0.005
Distal sensory conduction velocity	NS	<0.02	<0.01	<0.05
Ulnar nerve				
Conduction velocity of slower nerve fibres	<0.025	<0.05	<0.02	NS
Sensory conduction velocity	NS	NS	NS	<0.05
Posterior tibial nerve				
Maximum conduction velocity	NS	<0.05	<0.025	NS

NS = not significant.

It is clear from the table that the conduction velocity of the slower nerve fibres of the ulnar nerve was the most sensitive indicator. Even the subgroup with a maximum Pb–B level between 400 and 499 μg/l differed significantly from the control group. The subgroup whose maximum Pb–B level was between 500 and 599 μg/l showed four significant differences and the same differences also appeared in the 600–699 subgroup. The subgroup with a Pb–B level above 700 μg/l was not so consistent. This was probably owing to the fact that in this subgroup many subjects had been transferred to jobs with less exposure after they had been found to have high Pb–B levels in a health examination; the reactions may have been reversed to some degree.

In conclusion it can be stated that reduction in some conduction velocities (maximum conduction velocity, conduction velocity of the slower nerve fibres, sensory conduction velocity), especially of arm nerves, has been reported to occur at a Pb–B range of 200–700 μg/l. The no-effect level has not yet been clearly defined. Even on the basis of regression lines shown in Fig. 4 and 5, no clear threshold can be defined. On the other hand the significance tests place it in the 400–500 µg/l range. This refers to the highest Pb–B level ever recorded. However, it is not known which factor carries the higher risk—a high Pb–B level of short duration, or a moderately increased Pb–B level persisting for a longer period of time. More definite conclusions would require longitudinal studies of large populations whose exposure histories are fully documented. But it is extremely difficult to find enough subjects who have been both regularly monitored for a long time and who have never had high Pb–B levels. The lack of such subjects largely explains why so few studies have been published on the effects of low-level lead exposure on peripheral nervous functions.

Effects on the central nervous system. Although there is no doubt that heavy lead exposure causes encephalopathy (*2*), there is some controversy as to whether low intensities of exposure can cause subtle changes in central nervous functions. The following paragraphs focus on slight impairments of the central nervous system (CNS) in exposed workers; these impairments are considered to be more relevant than clinical encephalopathy from the point of view of defining health-based criteria for occupational exposure.

The toxic effects of lead on the central nervous system of infants and children are not very relevant in this context; however, from the point of view of preventing nervous damage in the fetus when pregnant women are exposed, paediatric experiences would be valuable. Unfortunately, however, there are very few, if any, exposure-effect and exposure-response data on Pb–B levels that may cause brain damage in children. Therefore, only a general assumption can be made that the developing CNS of infants, and especially of the fetus, is more sensitive to the toxic effects of lead than the CNS of adults (see also 3.2.7 below).

Clinically significant impairments of the central nervous system can be studied using a variety of neurological techniques, but for the study of subtle effects there is a narrower choice of methods. The most widely used methods for this purpose include different psychological tests and questionnaires.

So far four research groups have published results on the psychological performance of lead-exposed workers with Pb–B levels below 800 µg/l. All studies found some sort of impairment, but when the results are compared with the average performance of non-exposed normal individuals they are not consistent. This variation is probably at least partly due to the fact that one study relied on the maximum Pb–B level ever measured and the other three on current Pb–B levels.

In the last-mentioned studies the levels may not have been representative of lead exposure in previous months or years. Differences in the performance levels of the control groups used may have been another factor. Finding a control group that will correspond to the group to be examined with respect to primary performance level is indeed difficult. Therefore intragroup comparisons in general are more meaningful. It was seen that the results of the four research groups became more consistent when the performances, as measured by similar tests, were related to the Pb–B levels within the exposed groups.

Repko et al. (*69*), in a study of 316 workers with an average current Pb–B level of 630 μg/l (SD ± 217), found that the Pb–B level was significantly related to visual–motor functions, i.e., eye–hand coordination. In a later study comprising 42 storage battery workers, Repko et al (*8*) corroborated these findings. The range of the current Pb–B levels in these workers was 120–790 μg/l.

Hänninen et al. (*11*) found a relationship between lead in blood and impaired psychological performance among 49 workers whose highest Pb–B ever monitored had not exceeded 700 μg/l. The average Pb–B level was 320 μg/l at the examination (SD ± 11). The performances that were most affected by lead were those dependent on visual intelligence and visual-motor functions. The two tests that appeared to be the most sensitive indicators of lead-induced impairment were block design for visual intelligence, and the Santa Ana dexterity test for visual–motor ability.

Valciukas et al. (*13*) also found a relationship between the Pb–B levels and some psychological performances. They studied 99 secondary lead smelter workers whose *current* Pb–B levels were distributed as follows: 17% less than 400 μg/l, 61% between 400 and 590 μg/l, 20% between 600 and 790 μg/l, and 1% more than 800 μg/l. The four tests used in this study were all for visual intelligence and visual–motor function. Three of them yielded statistically significant differences between the exposed group and a control group comprising 25 steel workers. They also displayed statistically significant correlations with the Pb–B levels within the examined group. An even better correlation was found between the psychological performance and the zinc protoporphyrin IX (ZnPP) content of the erythrocytes, which may indicate that earlier, not current, Pb–B levels were more relevant.

Grandjean et al. (*12*) found impaired performance in a wide area of psychological function among 42 workers exposed to lead. The current Pb–B levels of the workers ranged from 120 to 820 μg/l with a mean value of 450 μg/l. Impaired performances covered both verbal and visual intelligence and memory functions; for most tests the decrease in performance was associated with indicators of lead exposure (Pb–B and ZnPP) within the exposed group, i.e., statistically significant correlations between Pb–B and a great number of tests were obtained. The wider range of psychological impairments found in this study, compared with those found in the other studies, may be explained on the basis of a higher average Pb–B value.

Effects on the central nervous system have also been studied by means of symptom questionnaires inquiring into subjective symptoms reflecting disturbances of sleep, mood, memory, attention, etc. Hänninen et al. (*10*) divided the 49 workers referred to above into two groups—those with maximum Pb–B levels below 500 μg/l and those with a level between 500 and 690 μg/l. Several symptoms were more prevalent in the higher exposure category. They included fatigue, sleep disturbances, forgetfulness, absentmindedness, restlessness, paraesthesia of the upper limbs, weakness of the lower limbs, and difficulties in walking in the dark. Corresponding results have been reported by Repko et al (*69*) and Lilis et al. (*9*), who found an excess of central nervous system impairment symptoms and pain in the muscles and joints of workers whose current Pb–B did not exceed 800 μg/l; the frequency of symptoms was related to the ZnPP level.

3.2.4 *Critical effects of lead exposure*

The Task Group on Metal Toxicity (*3*) defined the term "critical effect" to denote the earliest adverse (or undesirable) effect in an organ. Critical effects may be regarded in qualitative terms (e.g., degeneration of a nerve) or quantitative terms (e.g., a certain degree of protoporphyrin accumulation in the erythrocytes) and may or may not be of functional importance in relation to the health of the individual.

In this context the critical effects for lead are considered to be identical with the relevant effects defined on page 43 i.e., they comprise (1) disturbance of haem synthesis as indicated by concentrations of intermediates in blood and urine exceeding an agreed level and (2) subclinical central and peripheral neurological effects.

3.2.5 *Exposure-response relationships of the critical effects*

Haematological effects. When increases in protoporphyrin (PP) and ALA-U concentrations are considered as critical effects, one must be aware of the fact that the health significance of slight increases is not known. Exactly what intensity of an effect implies a health hazard is not known either, and variation between individuals must be presumed. This section will review the dose-response data available without taking any standpoint on their health significance, which will be discussed under 3.2.6.

The WHO Environmental Health Criteria publication on lead (*2*) contains exposure-response curves for Pb–B levels and the increase in erythrocyte PPs in males, females, and children, calculated by Zielhuis (*63*). In these calculations absolute values could not be used because the data were taken from authors who had used different methods for measuring erythrocytic PP. A cut-off level was calculated for PP and free erythrocyte protoporphyrin (FEP) levels at which about 95% of the subjects had a Pb–B concentration of less than 200 µg/l. Fig. 6 shows the percentage of subjects with raised FEP

Fig. 6. Exposure-response curves for Pb–B and the percentage of subjects with free erythrocyte protoporphyrin levels below the cut-off point

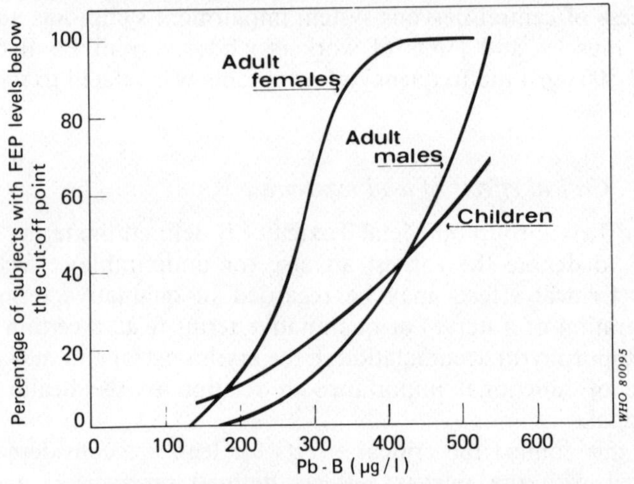

"Response" is taken to be the level of FEP that is not exceeded by 95% of the subjects whose Pb–B level does not exceed 200 µg/l (*63*)

levels in relation to increase in Pb–B concentration. In women the increase in FEP occurred at a lower Pb–B than in men. When these curves were plotted, the data for females were limited. However, later publications have corroborated this assumption (*52, 60*).

According to Zielhuis' calculations, the 5% response level for increase in erythrocytic PP occurred at a Pb–B concentration of about 200–250 μg/l for adult females and children and 250–300 μg/l for adult males. The 20% response level for adult males was about 400 μg/l. According to Roels et al. (*60*), there was a 20% response for females at a Pb–B level of 200 μg/l, a 50% response at 270 μg/l, and a 100% response at 440 μg/l. For males the corresponding levels were 280, 350 and 580 μg/l respectively. The exposure-response curves are shown in Fig. 7, which also shows exposure-response relationships for ALA-U. At a mean Pb–B level of 390 μg/l, 50% of the females had "abnormal" values while 100% showed "abnormal" values at a Pb–B

Fig. 7. Exposure-response curves for Pb–B and the percentage of subjects (men and women) with free erythrocyte protoporphyrin (FEP) and ALA-U levels below the cut-off point

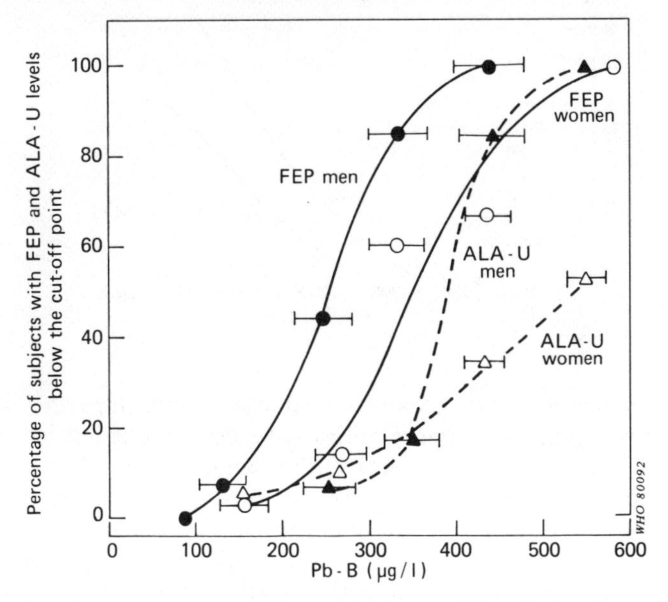

For women, the cut-off points for FEP and ALA-U were 826 μg/l of erythrocytes and 5.44 mg/g of creatinine respectively; for men, the cut-off points for FEP and ALA-U were 708 μg/l of erythrocytes and 4.76 mg/g of creatinine respectively (*60*)

level of 550 μg/l. The 50% response level was about 600 μg Pb/l of blood for the males; the data did not show the 100% response level.

The exposure-response relationship for ALA-U, based on calculations made by Zielhuis on male workers (*63*), has also been presented in the Environmental Health Criteria publication on lead (*2*). The results are presented in Fig. 8. According to these calculations, the 5% response level for ALA-U > 5 mg/l and > 10 mg/l occurred at Pb–B concentrations of about 300–400 μg/l and 400–500 μg/l, respectively. At Pb–B levels between 600 and 700 μg/l, 88% of the subjects had an ALA-U concentration above 5 mg/l and 50% had a value above 10 mg/l.

Fig. 8. Exposure-response curves for Pb–B and the percentage of subjects with ALA-U levels of ⩾5 mg/l and 10 mg/l (*63*)

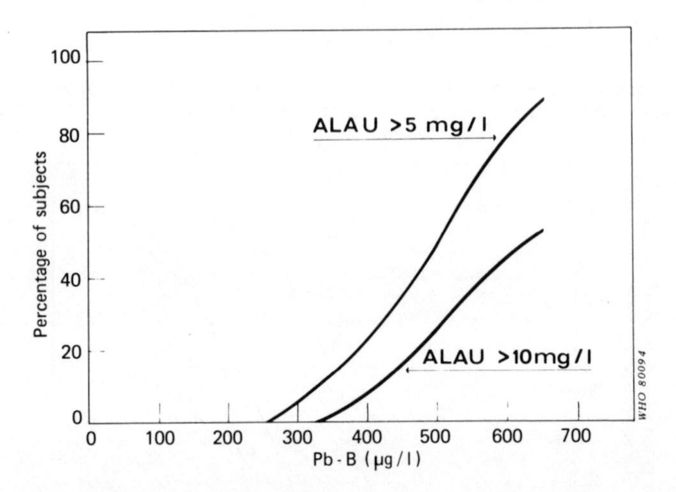

Neurological effects. Exposure-response relationships in relation to neurophysiological abnormalities have been presented in, or are possible to calculate from, a few studies only. The response level for nerve conduction velocities can be defined (1) as values that are lower than the lower normal limit of the laboratory in question; (2) as values that are lower than the lowest confidence limit, at the 95% confidence level, of an *ad hoc* reference population; and (3) sometimes as values that are lower than the normal limit reported in the literature.

Table 5 shows the exposure-response relationship for the maximum conduction velocity of the median nerve. Here 50 m/s has been considered as the normal lower limit, i.e., the cut-off point.

Table 5. Exposure-response relationships between the current Pb–B level and the maximum conduction velocity of the median nerve (*67*)

Pb–B category (μg/l)	No. of workers	No. of normal values	No. of abnormal values
up to 300	20	20	0
301–500	12	10	2
501–730	6	3	3

Although the numbers are small, there is a hint of a response level in the range of 300–500 μg/l, which becomes more convincing in the range of 510–730 μg/l.

In the study of Seppäläinen et al. (*68*), referred to earlier, the exposure-response relationship was displayed as a combination of pathological findings in nine different measurements of four nerves. This was expressed as a "conduction velocity score" corresponding to the number of abnormal conduction velocities for each lead worker. Abnormal values for each specific measurement were those that were below the mean −2 SD computed from the examination of a large number of people in the same laboratory. By definition a completely normal subject received a score of 0. The lead workers were divided into exposure categories on the basis of the maximum Pb–B levels ever measured, and 34 non-exposed industrial workers were used as the control group. The current mean Pb–B of the control was 110 μg/l. The results of this study are shown in Table 6.

As can be seen, there are two different measures of "response". When one pathological finding was used as the criterion, there was a suggestion of a response level below 500 μg/l. However, the difference in response rate between this subgroup and the control group was not statistically significant. But for all other groups significant differences were obtained. When the requirement was two or more pathological findings, the level was approximately 500–600 μg/l, and the response rate was fairly consistent in the higher exposure categories. These differences were not statistically significant, however. These findings should be compared with the exposure-effect data shown in Table 4, where a statistically significantly decreased mean value of the conduction velocity of slow nerve

Table 6. Conduction velocity scores (number of abnormal conduction velocities [a]) for lead workers in various exposure categories (based on the highest Pb–B level ever recorded) and the percentage of those with abnormal score (*68*)

Maximum Pb–B (μg/l)	N [b]	Score					Percentage with score ⩾1	Probability [c]	Percentage with score ⩾2	Probability [c]
		0	1	2	3	4				
Control (current mean Pb–B = 110)	34	33	–	1	–	–	3		3	
<400	3	2	1	–	–	–	..	NS [d]	..	NS
400–499	11	8	3	–	–	–	27	NS	0	NS
500–599	28	19	5	2	1	1	32	0.0046	14	NS
600–699	19	11	5	2	–	1	42	0.0012	16	NS
⩾700	17	8	7	1	1	–	53	0.00013	12	NS

[a] Abnormal conduction values are those equal to the mean value recorded in the control group minus twice the standard deviation.

[b] Number of examinees.

[c] Two-tailed *P*-values calculated from the hypergeometric distribution (Fisher's exact test) for fourfold frequency table (exposure category versus control).

[d] Not significant (P >0.05).

fibres of the ulnar nerve occurred already in the Pb–B subgroup of 400–490 μg/l. The seeming discrepancy may be explained by the rather strict criteria used for defining "response", i.e., lower than the mean −2 SD of the values found in a "normal" group.

Together the results of these two studies indicate with some certainty that nerve conduction impairment is induced at Pb–B levels of 500–600 μg/l and even the exposure range of 400–500 μg/l may be sufficient to cause impairment of nervous function. These studies should be confirmed in other laboratories because the number of subjects studied in these two studies was rather small. It should also be noted that Araki & Honma (*67*) related the conduction velocities to current Pb–B values, while Seppäläinen et al. (*68*) used the maximum Pb–B value ever measured as the exposure-indicator. However, in the latter study the time-weighted average Pb–B level and to some extent the current Pb–B level also yielded similar results.

Furthermore, it should be mentioned that there are some studies that do not show similar results. For example, Pavlev et al. (*70*) found no statistically significant difference between the nerve conduction velocities of the ulnar nerve of 32 lead workers as compared to 14 control subjects. The current average Pb–B level of the exposed group was 530 ± 160 μg/l and the range of exposure time was 2–37 months. However, this study cannot be accepted as being true negative investigation and hence contradictory of the positive results found by others because it suffers from at least the following five weaknesses: (1) small groups, (2) short follow-up, (3) insensitive methods of measurement, (4) questionable comparability of exposed and control groups, and (5) concurrent exposure to zinc, which is a well-known antagonist to lead toxicity (*71*). "Negative" results derived from such studies should be regarded as uninformative.

Electromyographical abnormalities have been found in several studies, especially in cases of clinical poisoning, but only a few of them relate the findings to Pb–B levels low enough to be of relevance in this context. From the data published by Seppäläinen & Hernberg (*66*) and Seppäläinen et al. (*65*), it can be seen that out of 32 patients with lead poisoning—but without any clinical neurological abnormalities—pathological findings (fibrillations and a diminished number of motor units in maximum contraction) occurred in 15 cases (47%) whereas 9 out of 26 (35%) of the group of lead workers referred to in Table 6 (maximum Pb–B ⩾ 700 μg/l) showed pathological findings. However, electromyograms were taken only in those cases which showed abnormal or borderline conduction velocities, and therefore this is a conservative estimate.

Abbritti et al. (*72*) found a similar pattern in electromyograms of 118 lead-exposed workers. Out of 59 subjects with current Pb–B levels between 410 and 800 μg/l (past Pb–B levels not reported), 30 subjects (51%) had pathological electromyograms, while 32 out of 49 subjects (65%) with Pb–B levels between 810 and 1200 μg/l, and all 8 of those with a Pb–B level of ⩾ 1210 μg/l had pathological findings. In this comparison only those with normal maximum conduction velocities were considered. When a combination of reduced maximum conduction velocity and abnormal electromyograms was considered, 2% of those with Pb–B levels between 410 and 800 μg/l and 12% of those with Pb–B levels between 810 and 1200 μg/l showed both findings. In this study no information was given of past exposure levels, but the proportion of pathological electromyogram findings tallies well with the studies by Seppäläinen et al. (*65, 68*) referred to above. The threshold level for pathological electromyogram findings cannot be precisly derived from these data; according to Seppäläinen et al. (*65*) it is somewhere below a Pb–B level of 700 µg/l, and according to Abbritti et al. (*72*) it lies somewhere between 400 and 800 µg/l.

Altogether the neurophysiological studies cited indicate that the minimum effect and response levels for any type of peripheral neurological effect are probably in the range of 400–500 µg/l and possibly even below 400 µg/l.

When the Pb–B level increases, the frequency of abnormal conduction velocities and probably also of pathological electromyogram findings increases. More severe neurological manifestations such as paralysis have not been reported to occur at Pb–B levels below 800 µg/l.

Of the published studies on psychological disturbances in lead workers, there is only one which provides dose-response data (*11*) (see page 60). The performances most affected by lead exposure were visual intelligence and visual-motor functions. The most sensitive indicators of impairment were the block design test for visual intelligence and the Santa Ana dexterity test for visual–motor ability. In the Santa Ana test "poor" (1 standard deviation below the "normal" average) results did not occur at all among a group of 22 lead workers whose Pb–B levels had not exceeded 500 µg/l. In another group of 27 workers whose maximum Pb–B level had been between 510–700 µg/l, "poor" results occurred in 7 cases (26%). Accordingly in this test, the Pb–B level at which no response could be detected was approximately 500 µg/l. Tests measuring other psychological functions did not display any differences within the group. The number of examinees in the study was too small to make any conclusions, but the results suggest that the no-response level for other lead-induced psychological disturbances lies above 700 µg/l.

3.2.6 *Conclusions*

According to Zielhuis' (*63*) calculations presented in section 3.2.5 above, the detected 5% response level for increase of protoporphyrin IX occurs at a Pb–B concentration of about 200 µg/l for children, 200–250 µg/l for adult females and 250–300 µg/l for adult males. According to similar calculations for ALA-U, the 5% response level for values in excess of 5 mg/l and 10 mg/l occurs at Pb–B concentrations of about 300–400 µg/l and 400–500 µg/l respectively. However, the health significance of such slight increases in the concentrations of haem intermediates is open to debate. Although these increases fulfil the criteria for a critical effect as defined in section 3.2.4, it is by no means clear whether or not a response defined as a slight increase above "normal" affects the function of the entire organism. It can be argued that slight increases can be well tolerated and that they therefore need not be considered when deriving a health-based occupational exposure limit. On the other hand, it is equally difficult to define any other cut-off point on the exposure-effect curve above which the increase should be considered harmful, and, inversely, below which it is considered to be without any health significance.

The earliest peripheral nervous effects begin to occur in the Pb–B range of 400–500 μg/l; in a few individuals they may occur below 400 μg/l. These effects consist of a slowing of the conduction velocities of the slow fibres of the ulnar nerve. In the Pb–B range of 500–600 μg/l other conduction velocities also become affected. Although it is probable that slight subclinical abnormalities in peripheral nervous function do not cause health impairment, changes of this type must be regarded more seriously than subclinical disturbances of the haem synthesis. The regeneration of the nervous system is slow, and as long as causative Pb–B levels persist it may well be that the regeneration that occurs is surpassed by continuing degeneration. Furthermore, since it is well known that heavy lead exposure ultimately causes paresis, slight subclinical changes might have to be regarded as precursors of more severe damage. Furthermore, it cannot be excluded that subclinical damage may occur concomitantly in the central nervous system although present methods fail to detect it. Support for this assumption can be found in the Finnish studies, where the same group of workers showed both peripheral nervous impairment and an excess of subjective symptoms as well as poor performance in some psychological tests (*11, 68*). For these reasons it may be prudent to consider any alteration in peripheral nervous function as adverse.

There are too few valid studies on psychological disturbances caused by low or moderate lead exposure in the adult. However, according to the scanty data available, some impairment in visual motor function and visual intelligence begins to occur at Pb–B levels above 500 μg/l. Even a suspicion of CNS damage warrants caution, and if future studies confirm the findings reported hitherto, such effects must be considered adverse.

Effects on reproduction and on the integrity of chromosomes have not been regarded as critical effects in this report, but they still warrant some comments. Quantitative data on chromosomal aberrations are scanty. A study by Nordenson et al. (*23*) suggests that an effect threshold for these changes is at a mean Pb–B level of about 250 μg/l; however, the value of these observations is weakened by failure to exclude concurrent exposure to other mutagenic metals. Although the final answer to the question whether or not chromosomal aberrations have health significance is still open, this finding must be viewed seriously against the background of known effects of higher doses of lead upon male and female fertility and congenital malformations (*2*). Hence it may be concluded that, although the results of studies on chromosomal aberrations in workers in the low-exposure range were equivocal, this finding should be regarded as adverse. At present, the data available are not convincing enough to provide a basis for a health criterion, though this situation may change in the very near future. On the basis of the above considerations and the data presented in previous sections, the following panorama of effects at various Pb–B levels can be drawn.

Pb–B ⩽ 300 μg/l. An increase in the erythrocytic protoporphyrin IX (PP) content in about 50% of females and about 15% of males is the only discernible relevant effect.

Pb–B 300–399 μg/l. About 90% of females and about 40% of males show an elevation of the erythrocytic PP. In males about 15% of the ALA-U values exceed 5 mg/l and some values exceed 10 mg/l. Approximately 40% of females excrete more than 5 mg ALA/l. There are suggestions of slight decrease in mean conduction velocities of peripheral nerves, although values below the normal lower range have not been demonstrated.

Pb–B 400–499 μg/l. The PP content of the erythrocytes exceeds "normal" values in all females and in more than 50% of males. The urinary level of ALA coproporphyrin (CP) is slightly elevated in a low proportion (< 20%) of lead-exposed males and in about 50% of females. The conduction velocity of the slow nerve fibres of the ulnar nerve is slightly decreased (this nerve is technically the easiest one to measure) and is probably below the normal lower limit in a proportion of cases.

Pb–B 500–599 μg/l. About 50% of subjects show a marked increase in the erythrocytic PP and all values are above "normal". The ALA-U and CP-U is elevated in about 50% of males and nearly 100% of females. About 10–20% of subjects show slowing of the conduction velocities of two or more nerves. Impairment of visual intelligence and visual–motor functions can be detected in 20–30% of subjects. According to one study, the frequency of subjective symptoms begins to increase.

Pb–B 600–699 μg/l: The ALA-U exceeds 10 μg/l in more than 50% of subjects; CP-U exceeds 100 μg/l; and the erythrocyte PP level exceeds 3000 μg/l in a majority of subjects. Slowing of the nerve conduction velocities shows a slightly higher frequency than in the preceding group. Pathological electromyograms occur in at least one-third of subjects. Psychological impairment and subjective symptoms occur probably at the same frequency as in the preceding category (the number of individuals studied has been too small to allow distinctions). Other effects, not discussed in this report, such as shortening of the erythrocyte life-span and lowering of the haemoglobin value also begin to appear.

3.2.7 *Effects of lead exposure on pregnant women*

The available data show that females and children are more sensitive than males to the toxic effects of lead on haem synthesis, at least with regard to inhibition of ferrochelatase, which results in an increased accumulation of PP in erythrocytes. These observations suggest that, for full protection, females require a lower health-based occupational exposure limit for Pb–B than males. Furthermore, in pregnant women protection of the developing embryo from injury to the central nervous system due to lead exposure is an important point to be considered. Because lead passes the placenta, and because placental blood has a Pb–B concentration proportional to maternal blood, there is no doubt that the fetus is exposed to much the same concentration of lead as the mother. Although no direct evidence is available on the effects of prenatal lead exposure on the CNS function in humans, it can be deduced that the developing brain is probably more vulnerable to the toxic actions of lead than the mature brain. Animal experiments also support this assumption (*47, 73–78*).

Because it is not known how early in the course of the pregnancy CNS damage can be caused by lead, and because elevated levels of lead in tissues persist for several months after the cessation of exposure, one cannot rely on self-reporting of pregnancies at an early stage to be effective in protecting the fetus. The only alternative is to apply more stringent health criteria for females in the fertile age, i.e., below 45–50 years, than for other categories of workers.

3.2.8 *Problems related to deriving health-based biological limits for populations that already have high non-occupational exposure*

There are areas in the world where the mean Pb–B level of the general population is so high that a proportion of the individuals will exceed the safe level recommended in this report. Irrespective of the criterion of health risk, no additional exposure of occupational origin can be permitted for persons close to or exceeding the recommended health-based biological limit for exposure. In such areas individuals showing low Pb–B levels should be preferred when choosing employees for work involving exposure to lead. This again emphasizes the importance of pre-employment screening for the Pb–B level.

This precaution should be particularly applied to women of child-bearing age. The lead exposure should be kept as low as possible and they should not be exposed to intensities that may cause their Pb–B levels to exceed the upper range for the general female population.

3.3 Research possibilities

Although lead is probably the most intensively studied occupational toxic agent, several questions regarding its toxicity remain. In this context, only gaps relevant to the recommendation of health-based occupational exposure limits and/or biological limits are discussed below.

3.3.1 *Relation between lead levels in blood and in air*

The Pb–B level is at present the best available indicator of personal exposure. However, since occupational exposure occurs mainly through lead in the workroom air, and the level of lead in air is the primary guideline for engineering purposes, further research is warranted on the following aspects of lead concentrations in air.

The effects of particle size, chemical composition (including solubility), and non-respirable particles on Pb–B levels is not clearly understood. At the individual level, some distinction must be made between non-occupational and occupational exposure; this can be achieved by studying new lead workers. Exactly how much exposure occurs from lack of personal hygiene i.e., exposure from contaminated fingers, cigarettes, and other non-food and food items, is not known. The protective effects of the use of respirators also need to be investigated.

3.3.2 *Exposure-effect and exposure-response relationship for nervous effects*

The effects of increases in erythrocyte PP on nervous function and the mechanisms underlying such effects are not clearly understood. The relationship between Pb–B level and neurophysiological disturbances also requires further investigation. It is also not known whether it is the Pb–B level *per se* or the increase in Pb–B level (above the level that may be found in the general population) due to occupational exposure that causes the neurological disturbances.

There is a dearth of data on exposure-effect and exposure-response of Pb–B levels with regard to subclinical impairment of CNS function as measured by psychological tests and neurophysiological methods.

Reversibility of nervous effects is still not very well understood. It should be studied in workers whose exposure has ceased owing to retirement or other reasons.

Further investigations are required on the no-response and no-effect levels for neurophysiological disturbances and on the diferences in susceptibility between males, females, growing embryos, and infants. Studies on children with known prenatal exposure will also be valuable.

Finally, better knowledge is required of the factors that increase or decrease the toxicity of lead with regard to CNS dysfunction.

3.3.3 *Mutagenic effects*

This is an area which has become particularly interesting during recent years because of the development of new methods for investigating mutagenic effects. While there is no doubt that high lead exposure causes a multitude of mutagenic effects (*2*), very few data exist on the effects of lower exposure intensities, i.e., those relevant to present occupational conditions. In this context, the studies that are required are the following:

— epidemiological studies on the carcinogenicity of lead;
— studies on the relationship between Pb–B levels and chromosomal aberrations utilizing sufficiently large populations representing a wide exposure range; and
— studies on the effects of lead on spermatogenesis, oogenesis, frequency of abortions, and malformations in both males and females; such studies should also focus on exposure-response relationships.

3.3.4 *Haematological effects*

In contrast to the problems referred to in sections 3.3.1, 3.3.2 and 3.3.3, there is a considerable body of knowledge on the relationship between Pb–B levels and haematological disturbances. Hence further research on haematological disturbances does not have the same priority. However, completion of existing exposure-effect and exposure-response data on Pb–B levels and erythrocyte protoporphyrin concentrations are still needed, particularly with respect to sex and age.

3.3.5 *Interactions*

Lead toxicity may be influenced by other environmental factors (synergistic, antagonistic, or additive), such as diet, deficiency states, concurrent diseases, and concurrent occupational exposures. The effects of these factors should be elucidated.

3.4 Recommendations

3.4.1 *Recommended health-based biological limit for Pb–B*

For the recommendation of the health-based biological limit for Pb–B it can be assumed that central and peripheral neurological effects are more relevant than a slight increase of haem metabolism intermediates. However, strictly speaking, even the inhibition of haem synthesis, as manifested by an increase of intermediates, can be regarded as adverse.

Irrespective of the type of effect used as the basis for recommending exposure limits, some arbitrariness cannot be avoided because there are difficulties in defining no-adverse-effect levels, as is evident from the curves shown in Fig. 1, 3, 4 and 5. Furthermore, one should consider the differences in sensitivity between males and females with respect to effects on the haem synthesis, and, in addition, the special considerations that arise from the demand to protect the offspring and females in the fertile age range.

Taking into account the insufficiently confirmed data on the minimum-adverse-effect level for the nervous system, a health-based biological exposure limit for Pb–B of 400 μg/l is recommended for males and for females over the reproductive age. This limit is based on the adverse effects of lead on the haematopoietic and peripheral nervous system at the Pb–B range of 400–490 μg/l. The data from Tables 4 and 6 also show more pronounced effects of lead at Pb–B levels of 500 μg/l and above, at which levels effects on the CNS also begin to be observed.

According to present knowledge, Pb–B levels of less than 400 μg/l do not cause adverse neurological effects, although it must be admitted that a no-effect level is impossible to define from the available data. Slight increases in haem synthesis intermediates, especially in the females, are not prevented, however. But in view of the uncertainty of the health significance of such mild effects, there are not at present strong enough reasons to prompt the recommendation of a limit that would entirely prevent their appearance.

The Pb–B level of females in the reproductive age range should be kept as low as possible and, to protect the fetus, should not exceed 300 µg/l. This level will also prevent significant effects on haem synthesis. In practice this recommendation may imply restrictions for women of child-bearing age to work in areas where exposure may cause Pb–B levels to exceed the 300 µg/l limit.

3.4.2 *Recommended health-based biological limit for lead in urine*

Urine analysis is not recommended for health monitoring; hence no recommendation is given for the concentration of lead in urine.

3.4.3 *Health-based recommendations for PP and ALA-U*

Since Pb–B measurements are impracticable in some countries, recommendations for PP and ALA-U are required. It should be stressed, however, that these substances provide only measures of *effect* and that they only indirectly reflect exposure. It should also be stressed that, because of the rather wide confidence limits for their correlations with Pb–B, values considerably lower than those corresponding to the average regression line must be recommended. Otherwise some of the Pb–B concentrations at a given ALA-U or PP value may reach levels that can cause nervous impairment. Because of the diversity of the methods used, relative values are recommended here.

For ALA-U, no increase over the laboratory's "normal" upper limit (e.g., mean + 2 SD) should occur. In the case of PP, that level may be raised by up to 50%. The "normal" refers in both cases to values derived from a general adult population whose average Pb–B level does not exceed 200 µg/l, taking into account the difference in susceptibility in males and females.

3.4.4 *Recommended health-based exposure limit for concentration of lead in air*

As discussed previously, the correlations between Pb–B and lead in air are so weak that recommended Pb–B limits cannot be translated directly into limits for lead in air. Furthermore, the data on direct relationship between lead in air and effects are insufficient for the purpose of deriving the recommendation for air lead without using the Pb–B as an intermediate step.

In the extrapolation of these data one must recognize that, while the Pb–B value is the primary criterion for preventing toxic effects in the exposed workers, the concentration of lead in air is the primary guideline for engineering purposes. If the level of lead in air is kept low enough, the Pb–B level will not exceed the safe upper limit, if other sources of exposure can be eliminated. Since the contribution to the Pb–B level of non-occupational sources varies between subjects and between geographical areas, its relative share is increased when the Pb–B recommendation is low, and since any relation between lead in air and Pb–B levels will also depend upon the particle size and chemical form of lead, it is impossible to recommend a health-based occupational exposure limit for lead in air that would by itself ensure full compliance with the Pb–B level of 400 μg/l recommended above. Therefore, the evaluation of excessive absorption *primarily depends on biological monitoring.* However, as discussed in section 3.3.1, an average increase of 5 μg Pb/l blood can be predicted from each average increase of 1 $\mu g/m^3$ (40 h per week average exposure) of lead in air up to a Pb–B level of 500 μg/l. Hence, for populations with an average basic Pb–B level of 250 μg/l, the value of the level of lead in air should not exceed 30 $\mu g/m^3$, while 60 $\mu g/m^3$ can be tolerated for populations with basic average Pb–B levels of 100 μg/l. On this reasoning, a range of 30–60 $\mu g/m^3$ is recommended for lead in air. The application of ranges for air lead hence presupposes the use of both pre-employment screening and regular follow-up of Pb–B levels. In addition, close attention should be paid to matters of personal hygiene.

Regular biological monitoring and adherence to the air lead limit appropriate for each population can prevent the occurrence of the adverse effects.

The above recommendations are applicable only when the quality control of the laboratory methods is satisfactory.

REFERENCES

1. TASK GROUP ON METAL ACCUMULATION. *Environmental physiology and biochemistry*, **3**: 65–107 (1973).
2. WORLD HEALTH ORGANIZATION. *Environmental health criteria 3. Lead*, Geneva, 1977.
3. NORDBERG, G. F., ED. *Effects and dose-response relationships of toxic metals. Proceedings of an international meeting organized by the Subcommittee on the Toxicology of Metals of the Permanent Commission and International Association on Occupational Health, Tokyo, 18–23 November 1974*, Amsterdam, Elsevier, 1976.
4. HAEGER-ARONSEN, B. *British journal of industrial medicine*, **28**: 52 (1971).
5. NORDMAN, C. H. *Environmental lead exposure in Finland. A study on selected population groups.* Doctoral thesis, University of Helsinki, 1975.
6. RYU, J. E. ET AL. *Journal of pediatrics*, **93**: 476 (1978).
7. HERNBERG, S. Lead. In: Zenz, C., ed. *Occupational medicine; principles and practical applications.* Chicago, Yearbook Medical Publishers, 1975.
8. REPKO, J. D. ET AL. *The effects of inorganic lead on behavioural and neurologic function. Final report.* Washington, DC, US Department of Health, Education and Welfare, 1978 (Publication No. 78–128).
9. LILIS, R. ET AL. *Archives of environmental health*, **32**: 256 (1977).
10. HÄNNINEN, H. ET AL. *Neurotoxicology*, **1**: 313–332 (1979).
11. HÄNNINEN, H. ET AL. *Journal of occupational medicine*, **20**: 683 (1978).
12. GRANDJEAN, P. ET AL. *Scandinavian journal of work, environment and health*, **4**: 295 (1978).
13. VALCIUKAS, J. A. ET AL. *International archives of occupational and environmental health*, **41**: 217 (1978).
14. NEEDLEMAN, H. L. ET AL. *New England journal of medicine*, **300**: 689 (1979).
15. BUCHTAL, F. & BEHSE, F. *British journal of industrial medicine*, **36**: 135–147 (1979).
16. WEEDEN, R. P. ET AL. *Archives of international medicine*, **139**: 53 (1979).
17. BAKER, E. L. ET AL. *British journal of industrial medicine*, **36**: 314–322 (1979).
18. FREEMAN, R. *Archives of disease in childhood*, **40**: 389 (1965).
19. MYERSON, R. M. & EISENHAUER, J. H. *American journal of cardiology*, **11**: 409 (1963).
20. SILVER, W. & RODRIQUEZ-TORRES, R. *Pediatrics*, **41**: 1124 (1968).
21. DINGWALL-FORDYCE, I. & LANE, R. E. *British journal of industrial medicine*, **20**: 313 (1963).
22. SARTO, F. ET AL. *Medicina del lavoro*, **69**: 172 (1978) (in Italian).
23. NORDENSON, I. ET AL. *Hereditas*, **88**: 263 (1978).

24. Sirover, M. A. & Loeb, L. A. *Science*, **194**: 1434 (1976).

25. International Agency for Research on Cancer. *Evaluation of the carcinogenic risk of chemicals to man.* Volume 1, Lyons, 1972.

26. Cooper, W. C. Mortality of workers in lead production facilities and lead battery plants. In: *Abstracts of the XIX International Congress on Occupational Health*, Dubrovnik, Yugoslavia, 25–30 September, 1978. Institute for Medical Research and Occupational Health, Zagreb.

27. Subyanove, J. P. *Gigiena truda i professional'nye zabolevanija*, **12**: 24 (1978) (in Russian).

28. Federal Register, 14 November, 1978, *Part IV. Occupational exposure to lead. Final Standard.* Washington, DC, Department of Labor, Occupational Safety and Health Administration, 1978.

29. Tola, S. & Nordman, C. H. *Environmental research*, **13**: 250 (1977).

30. Williams, M. K. et al. *British journal of industrial medicine*, **26**: 202 (1969).

31. Williams, M. K. *International archives of occupational and environmental health*, **41**: 151 (1978).

32. National Institute of Occupational Safety and Health. *Health hazard evaluation determination report No. HE 77–28–552.* Washington, DC, US Department of Health, Education and Welfare, 1978.

33. Coulston, F. et al. *The effect of continuous exposure to airborne lead. 2. Exposure of man to particulate lead at a level of 10.9 μg/m³. Final Report to the US Environmental Protection Agency.* Research Triangle Park, NC, 1972.

34. Coulston, F. et al. *The effect of continuous exposure to airborne lead. 4. Exposure of man to particulate lead at a level of 3.2 μg/m³. Final Report to the US Environmental Protection Agency.* Research Triangle Park, NC, 1972.

35. Rabinowitz, M. B. *Science*, **182**: 725 (1973).

36. Rabinowitz, M. et al. *Environmental health perspectives*, **7**: 145 (1974).

37. Rabinowitz, M. et al. *Journal of laboratory and clinical medicine.* **90**: 228 (1977).

38. Bridbord, K. *Preventive medicine*, **7**: 311 (1978).

39. Noweir, M. H. et al. *Bulletin of the High Institute of Public Health (Alexandria)*, **5**: No. 1, 173 (1975).

40. Berlin, A. et al. Intercomparison programme on the analysis of lead, cadmium and mercury in biological fluids. In: *Proceedings of CEC-EPA-WHO International Symposium: Recent advances in the assessment of the health effects of environmental pollution*, Paris, 24–28 June, Luxembourg, Commission of the European Communities, 1974.

41. Donovan, D. T. et al. *Archives of environmental health*, **23**: 111 (1971).

42. Lamola, A. A. & Yamane, T. *Science*, **186**: 936 (1974).

43. Blumberg, W. E. et al. *Journal of laboratory and clinical medicine*, **89**: 712 (1977).

44. SASSA, S. ET AL. *Medicina del lavoro*, **69**: 172 (1978) (in Italian).

45. PIOMELLI, S. ET AL. *Pediatrics*, **51**: 254 (1973).

46. STUIK, E. J. *Internationales Archiv für Arbeitsmedizin*, **33**: 83 (1974).

47. CARSON, T. L. ET AL. *Archives of environmental health*, **29**: 154 (1974).

48. ROELS, H. A. ET AL. *Internationales Archiv für Arbeitsmedizin*, **34**: 97 (1975).

49. ALESSIO, L. ET AL. *International archives of occupational and environmental health*, **38**: 77 (1976).

50. ALESSIO, L. ET AL. *International archives of occupational and environmental health*, **37**: 73 (1976).

51. ALESSIO, L. ET AL. *Medicina del lavoro*, **69**: 563 (1978) (in Italian).

52. ALESSIO, L. ET AL. *International archives of occupational and environmental health*, **40**: 283 (1977).

53. JOSELOW, M. M. & FLORES, J. *American Industrial Hygiene Association journal*, **38**: 63 (1977).

54. HAEGER-ARONSEN, B. & SCHÜTZ, A. *Läkartidningen*, **75**: 3427 (1978) (in Swedish).

55. BALOH, R. W. *Archives of environmental health*, **28**: 198 (1974).

56. HOTZ, P. ET AL. *Sozial und präventiv Medizin*, **22**: 121 (1977) (in French).

57. TOMOKUNI, K. & OGATA, M. *Archives of toxicology*, **35**: 239 (1976).

58. ZIELHUIS, R. L. *International archives of occupational and environmental health*, **39**: 59 (1977).

59. ZIELHUIS, R. L. ET AL. *International archives of occupational and environmental health*, **42**: 231 (1979).

60. ROELS, H. ET AL. Sex and chelation therapy in relationship to erythrocyte protoporphyrin and urinary δ-aminolevulinic acid in lead exposed workers. *Journal of occupational medicine*, (in press).

61. KAHN, H. A. Early detection of toxic substances on the organism and the problem of individual susceptibility to toxic substances. In: Hernberg, S. & Kahn, H. A. ed. *Proceedings "Effects of early action of toxic substances on the organism*, Institute of Experimental and Clinical Medicine of the Ministry of Health of the Estonian S.S.R., Tallinn, and Institute of Occupational Health, Helsinki, Tallinn, 1977, pp. 9–17.

62. KAHN, H. A. & MERE, A. T. *Gigiena truda i professional'nye zabolevanija*, **12**: 28 (1978).

63. ZIELHUIS, R. L. *International archives of occupational health*, **35**: 1 (1975).

64. SHIMAITIS, R. S. & TSUNES, E. P. *Gigiena truda i professional'nye zabolevanija*, **12**: 50 (1977) (in Russian).

65. SEPPÄLÄINEN, A. M. ET AL. *Archives of environmental health*, **30**: 180, 1975.

66. SEPPÄLÄINEN, A. M. & HERNBERG, S. *British journal of industrial medicine*, **29**: 443 (1972).

67. ARAKI, S. & HONMA, T. *Scandinavian journal of work environment and health*, **4**: 225 (1976).

68. SEPPÄLÄINEN, A. M. ET AL. *Neurotoxicology*, **1**: 333–347 (1979).

69. REPKO, J. D. ET AL. *Final report on the behavioural effects of occupational exposure to lead. Interim Technical Report No. ITR–74–27.* US Department of Health, Education and Welfare, Cincinnati, OH, February 1974 (revised March 1976).

70. PAVLEV, P.-E. ET AL. *International archives of occupational and environmental health*, **43**: 37 (1979).

71. DUTKIEWICZ, B. ET AL. *International archives of occupational and environmental health*, **42**: 341 (1979).

72. ABBRITTI, G. ET AL. *Medicina del lavoro*, **68**: 412 (1979) (in Italian).

73. SILBERGELD, E. K. & GOLDBERG, A. M. *Life science*, **13**: 1275 (1973).

74. MICHAELSON, I. A. *Toxicology and applied pharmacology*, **26**: 539 (1973).

75. BRADY, K. ET AL. *Pharmacology biochemistry and behaviour*, **3**: 561 (1975).

76. REITER, L. W. ET AL. *Environmental health perspectives*, **12**: 119 (1975).

77. BROWN, D. R. *Toxicology and applied pharmacology*, **32**: 628 (1975).

78. SOBOTKA, T. J. *Toxicology*, **5**: 175 (1975).

79. LAUWERYS, R. ET AL. *Environmental research*, **15**: 278 (1978).

4. MANGANESE

4.1 Summary of metabolism and toxicity

4.1.1 *Metabolism*

In occupational exposure manganese is absorbed mainly through the lungs. Manganese diffuses into the bloodstream only when the inhaled particles are small enough to reach the alveoli—the so-called "respirable" particles. Larger particles are cleared from the respiratory tract and eventually swallowed. This is owing to the fact that only some manganese salts (e.g., permanganate, sulfate, chloride, and nitrate) are water-soluble; manganese dioxide, the major component of manganese ores, and the compounds used in metal refining or produced as by-products are all practically insoluble in water. This property of the compound plays a crucial role in their absorbance through the lungs.

Manganese may also be absorbed through the gastrointestinal tract with food and water. Only about 3% of the ingested manganese is absorbed. Little is known about the mechanism of absorption but studies indicate that the amount of iron in the body influences the mode of absorption. When the iron content of the body is high, manganese is absorbed in the intestines by diffusion, and when the body is iron-deficient the absorption takes place by active transport in the duodenum and jejunum. Absorption of manganese through the skin is negligible.

The absorbed manganese leaves the bloodstream quickly and concentrates primarily in the liver. The kinetic patterns of blood clearance and liver uptake of manganese are similar, indicating that these two manganese pools rapidly enter into equilibrium. The excess metal may be distributed to other tissues but most of it is discharged into the bile. Manganese concentrated in the liver is conjugated with the bile salts. Manganese distributed in other tissues, being in a highly mobile and dynamic state, enters the mitochondria. Some manganese also enters the nuclei; the accumulation of manganese there occurs at a slower rate than the clearance of manganese from the bloodstream. Manganese passes the blood-brain barrier, and it can also enter the placenta.

Generally, organs and tissues do not accumulate large quantities of manganese. The manganese content is higher in tissues rich in mitochondria. In addition to the brain, the pituitary gland, pancreas, liver, kidneys, small intestine, lungs, and bones have a relatively high concentration of manganese. It has been found that hair also accumulates manganese, particularly dark hair. Higher concentrations of manganese are also associated with other pigmented portions of the body, including the retina, pigmented conjuctiva, and dark skin. The storage capacity of the liver for manganese is limited to about 1–1.3 mg/kg (wet weight). The accumulation of manganese in different organs including the lungs does not increase with age.

Manganese is an essential element for human beings. It is a metalloprotein component of some enzymes such as pyruvate decarboxylase and probably liver arginase. Several other enzymes such as prolidase and succinic dehydrogenase require manganese exclusively as an activator. In animals it is essential for normal bone formation and synthesis of chondroitin sulfate.

It is estimated that the total manganese body burden for a man of 70 kg is about 10–20 mg.

Manganese has a rather short biological half-life. In an experiment with isotope-labelled manganese dichloride in volunteers the biological half-life of the fast component for the whole body was

4 days. Some 60–65% of intravenously injected manganese was eliminated with an average half-life of 39 days, although in one subject 90% of manganese was eliminated during this time. Studies in animals demonstrated that the higher the concentration of manganese the animals were given, the shorter the biological half-life. Studies on the chemobiokinetics of manganese suggest that the biological half-life for manganese in the brain is considerably longer than for the whole body.

Manganese is mainly excreted with the bile into the alimentary canal from where it is almost entirely excreted with the faeces. Excretion also occurs via the pancreatic juice, the duodenum, jejunum, and to a lesser extent the terminal ileum. Manganese excreted with the bile enters the intestine where it is partly reabsorbed. It seems that bile excretion is the main regulatory mechanism accounting for the relative stability of manganese content in the tissues. Only a small amount of manganese (about 0.1–1.3% of the daily intake) is excreted in urine.

4.1.2 *Concentrations of manganese in biological materials as indicators of exposure*

Manganese has been measured in blood, urine, stools, and hair. Estimation of the degree of manganese exposure by means of manganese concentrations in urine and blood did not prove to be of great value. Urinary manganese may increase in manganese-exposed workers. While Tanaka & Lieben (*1*) and other investigators have shown that urinary manganese mean concentrations correlate roughly with mean air concentrations of manganese, in individual cases the correlation is poor.

The average manganese blood level in workers exposed to the substance seems to be of the same order as that in non-exposed persons. However, there are some observations indicating that heavy exposure to manganese may increase the level of manganese in the blood. In a recent study (*2*) it was found that workers exposed to about 1 mg of manganese dust per m^3 of air during a period of 1–10 years had blood levels of manganese between 11–16 µg/l compared to a mean level of 10 µg/l in non-exposed persons.

The determination of manganese in faeces has also been recommended as a group test for the evaluation of the level of occupational exposure to manganese. In a group of 115 workers in an electrode factory in Czechoslovakia, where manganese concentrations in air were much higher than the maximum allowable concentrations of 2 mg/m^3, the mean manganese value detected in faeces was 62.3 mg/kg wet weight. Values higher than 100 mg/kg wet weight of faeces were infrequent (*3*).

In another group of 4 workers from a manganese ore crushing plant, where the manganese concentration in air was up to 15 times higher than the value of 2 mg/m^3, manganese in urine and faeces was monitored for 5 days; during the last 4 days of monitoring the workers had no occupational exposure (*4*). Manganese in urine was seen to have increased in all exposed workers and a mean value of 18 μg/24 h was recorded compared with mean value of 12.8 μg/24 h in a group of non-exposed subjects. The mean excretion of manganese in faeces was 12.3 mg/24 h compared with 2.4 mg/24 h in the control group.

Attempts have also been made to measure the concentration of manganese in the hair of exposed workers. There is some evidence that manganese may concentrate in hair following exposure to high manganese levels in air (*5*). It seems to be concentrated in the melanin-containing pigmented granules, and hence a very low concentration is found in white hair and fingernails which are not pigmented (*6*).

4.1.3 *Normal concentrations of manganese in biological materials*

"Normal" mean values of manganese for human blood vary between 0.84 and 150 μg/l of whole blood. Diurnal and seasonal variations have also been observed, but in general, the manganese level in blood, as in tissue, is more or less constant. It has been suggested (*7*) that manganese in blood may be influenced by lead exposure. There is a tendency for manganese in blood to increase with a rise in blood-lead level. The manganese content of erythrocytes is higher than that of plasma. The manganese content of plasma and serum in people without occupational exposure to this metal has been found to be in the range of 1.83–3.1 μg/l and 0.36–2.97 μg/l respectively. The manganese in human serum is selectively and almost totally bound in the trivalent form by a β_1-globulin. In erythrocytes a firmly bound manganese compound has been observed which is probably a manganese porphyrin.

The mean concentration of manganese in the urine of non-exposed persons is usually estimated to be 1–8 μg/l, but values of up to 21 μg/l have been also reported.

Recent studies (*8, 9*) suggest that values for manganese in blood in non-exposed persons do not exceed 20 μg/l and in urine 2 μg/l. These data have been arrived at both by flameless atomic absorption and by neutron activation analysis.

A manganese concentration of 60 mg/kg of faeces has been suggested as indicative of occupational exposure to manganese. The manganese content in hair is normally below 4 mg/kg (*10*). Other biological materials that could be used to monitor manganese exposure have not yet been identified.

4.1.4 *Toxic effects*

Effects on the central nervous system. The first sign of manganese poisoning is impaired mental capacity. It may result from exposure to high concentrations of manganese dusts or fumes for only a few months or even less. Usually it appears after a prolonged exposure of two or more years. Disorders of the central nervous system are accompanied by many other symptoms. In its course, the disease can be divided into three phases:

(1) a subclinical stage with generally vague symptomatology;

(2) an early clinical stage in which psychic or neurological symptoms and signs predominate and include acute psychomotor disturbances, dysarthria, and disturbance of gait; and

(3) a fully developed stage associated with manic or depressive psychosis and parkinsonism.

Well-established manganese poisoning is a disease characterized by permanent disability. If the worker is removed from exposure shortly after the onset of neurological symptoms (i.e., prior to the fully developed stage of manganism) many of the symptoms and signs will disappear. However, there may remain some residual disturbances, particularly in speech and gait.

It is recognized that both parkinsonism and chronic manganese poisoning result from an unantagonized cholinergic excitation of the corpus striatum secondary to decreased dopaminergic inhibitory input following a lesion of the nigrostriatal tract. Autopsy of chronic manganese poisoning cases has shown that lesions of the central nervous system are most severe in the striatum and pallidum but may also be found in the substantia nigra. A post-mortem analysis of one case showed a reduced concentration of dopamine.

The many animal studies carried out on the neurotoxic effects of manganese provide strong evidence that the metabolic abnormalities in the brain and the destruction of "dopaminergic" neurons caused by manganese are very similar to those occurring in parkinsonism. Clinical and biochemical abnormalities may be evident before there is any microscopic destruction of nerve cells. This may be particularly relevant to the early psychotic manifestations in human manganism, which could also be attributed to a disturbance in amine metabolism.

An increase in the toxicity of manganese has been observed with concurrent exposure to certain chemicals, such as carbon monoxide, copper, and lead. Some studies indicate that exposure to manganese combined with vibrations or X-rays may also increase the toxic effect of manganese (*11, 12*).

In connection with the toxic effects of manganese a marked individual susceptibility has been observed. While some workers may be exposed for many years without any adverse effect, others develop manganism quickly. Factors possibly influencing individual susceptibility are alcoholism, chronic infections, and liver and kidney dysfunction. A high degree of individual susceptibility to chronic manganese poisoning has been also attributed to nutritional deficiencies, particularly to anaemia caused by iron deficiency. It appears that calcium can influence manganese metabolism by affecting the retention of absorbed manganese as well as by affecting its absorption. The susceptibility may be related to individual variations in the excretory capacity of the liver and kidneys, which may lead to accumulation of manganese to toxic levels. However, there are still many unsolved problems.

Daily dietary intake of manganese depends on the amount of unrefined cereals, nuts, leafy vegetables, and tea consumed by the individual. Tea-leaves contain far more manganese than any other food item. Thus, differences in dietary habits may be connected with the degree of exposure and may explain to some extent the differences in individual susceptibility to manganese exposure.

Effects on the lungs. In addition to the neurotoxic effects, an increased incidence of pneumonia has been reported after short-term exposure to high levels of manganese in air. The possible toxic effect of manganese on the lungs was overshadowed by its action on the central nervous system, particularly during earlier years. While the relationship between pneumonia and manganese exposure has been identified, some investigators do not accept it. For example, no findings suggesting that there is excess pneumonia in manganese workers have come from the USA or Canada. Those who support "manganese pneumonia" say that if the metal does not act as a direct etiological agent it probably acts as a predisposing factor. Manganese may interfere with some immunological mechanism rendering the individual more susceptible to infections.

Animal and other experimental studies (including cell culture studies) have confirmed the effect of manganese on the lungs. This relates to acute pneumonitis produced by direct intratracheal administration of manganese dioxide dust or solutions of manganese chloride. Inflammatory changes in lung tissue were also observed when animals inhaled manganese dioxide dust in a concentration of 0.3–0.7 mg/m^3. An increased "irritation score" has been observed in combined exposure to manganese and other irritants; an inhibitory action of manganese on alveolar macrophages has also been found.

In a recent animal study (*13*) it was shown that a primary inflammatory reaction may occur in the respiratory tract after exposure to manganese dioxide in the absence of pathogenic bacteria. This reaction, which is of a limited duration after a period of acute exposure, is related to the property of manganese dioxide dust to induce pathological reactions via the cellular elements normally present in the lung. In the author's opinion, the reaction pattern also explains the ineffectiveness of the usual antibiotics used for pneumonia treatment in the acute phase after exposure to manganese dioxide. Animal experiments also showed that the pulmonary reaction after exposure to manganese dioxide is more pronounced in lungs challenged with bacteria. It has been suggested that exposure to manganese dioxide causes a decrease in the resistance to respiratory infections. Findings that bacterial clearance is reduced after exposure to manganese dioxide support these observations. Consequently, some earlier cases of manganese pneumonia might have had a bacterial (or viral) genesis, particularly in groups of workers whose conditions were such that exposure to manganese dioxide was low, but there was a high risk of airborne infections.

Apart from the increased occurrence of pneumonia, an increased incidence of acute bronchitis in manganese-exposed workers has also been observed. It has also been suggested that long-term manganese-exposure may contribute to the development of chronic non-specific lung diseases particularly if the exposure is combined with smoking (*14*). A higher incidence of bronchial asthma in manganese miners than in a control population group has been reported (*15*).

Other effects. A hypotonic effect of manganese on the systolic blood pressure, with no such effect on diastolic pressure, has been observed (*16*). This observation has been supported by some previous data obtained from animal experiments which indicated that manganese can induce a decrease in blood pressure.

One study indicates that wives of workers exposed to manganese showed an increased rate of spontaneous abortions and stillbirths (*17*). The rate of spontaneous abortions appeared to increase also with increase in duration of exposure of their husbands. This study did not mention the wives' occupations.

Data on the effect of manganese on human blood are conflicting. According to some reports, long-term exposure to manganese increases haemoglobin values and erythrocyte counts; lymphocytosis and a decrease in the number of polymorphonuclear cells have also been observed.

Some other observations indicate that the toxic effects of manganese may include: dysproteinaemia with hypoalbuninaemia and hyper-beta-globulinaemia; increased bilirubin level; increased transaminase activity; reduced lactic dehydrogenase activity; reduced magnesium and increased calcium levels in blood serum; and reduced haemoglobin and glutathione levels in erythrocytes. In cases of mild intoxication, the level of serum adenosine deaminase may be elevated. In a number of patients with manganese poisoning a diminished excretion of the 17-ketosteroids was also noticed. In animals given excessive doses of manganese, nephritis with proteinuria, casts, and haematuria has been seen. Intravenous injections of 0.7–2 mg/kg body weight depressed the thyroid activity in rats. Disturbances of sex function and testicular changes affecting germinal activity after exposure to permanganate and to certain other manganese compounds have been noted. Intravenous administration of 3.5 mg of manganese chloride per kg of body weight to rabbits has been reported to produce histochemically detectable alterations in the testes confirmed by decreases in lipoamide reductase (NADH), succinate dehydrogenase, and glucose-6-phosphate dehydrogenase activity.

Mutagenic effects. No information is available on mutagenic or chromosomal abnormalities in humans due to manganese exposure. However, from animal experiments there are some data on increased incidence of chromosome aberrations in bone-marrow cells (30.9% compared to 8.5% in control rats) after administration of manganese chloride (*18*). Manganese chloride is also reported to be mutagenic in some microorganisms (some bacteria and yeasts). Several studies have shown that manganese inhibits and decreases the fidelity of DNA synthesis *in vitro* (*19, 20*). Using *Escherichia coli* RNA polymerase with calf thymus DNA (not dependent on sigma factor) and phage T4 DNA (sigma-factor-dependent) as templates, Hoffman & Niyogi (*21*) showed that manganese, in addition to decreasing the fidelity of DNA synthesis, stimulated chain initiation of RNA synthesis at manganese concentrations that inhibit overall RNA synthesis. Studying the error frequency of *in vitro* DNA synthesis, Weymouth & Loeb (*22*) showed that substitution of Mn^{2+} for Mg^{2+} and unequal concentrations of deoxynucleoside triphosphate substrates raised the mutation frequency to values greater than 1 in 1000.

One study shows that intraperitoneal administration of manganese(II) sulfate to mice over a period of 30 weeks increased the incidence of lung tumours. This effect was produced by 15 injections at a dose level of 10 mg/kg body weight (*23*).

Conclusions. Not all the toxic effects discussed above can be regarded as relevant in the assessment of the health risk in occupational exposure. The Group regarded the effects on the central nervous system and lungs as critical in making decisions on recommended health-based occupational exposure limits.

Other effects (e.g., decreased blood pressure, dysproteinaemia, and reproductive disturbances) cannot be considered as conclusive, either because they are non-specific and there are probably other factors involved, or because they are based on animal studies and there is no evidence of these effects occurring in humans (e.g., carcinogenic and mutagenic effects, depressed thyroid activity, and nephritis).

4.2 Relationship between exposure and health effects

4.2.1 *Exposure indicators*

Manganese in air. Information on manganese concentrations in workroom air is usually limited. Moreover, there is usually poor correlation between the measured mean manganese concentrations at workplaces and the degree of the disorders. This may be partly because the concentrations of manganese in the workroom air vary with time, and the results of a few measurements do not reflect the actual exposure. The effect depends very much on the particle size of the manganese aerosol. Prolonged exposure to relatively high concentrations of manganese dioxide dust as experienced by pneumatic drillers is associated with extensive impairment of the central nervous system. In contrast, a considerably lower exposure to manganese dioxide fumes may produce the same CNS impairment symptoms in those who melt the metal.

Considering the dependence of deposition sites and solubility rates on particle size and the resulting difference in toxic effects between manganese dioxide dust and fumes, it has been suggested that the health-based occupational exposure limits should reflect explicitly the key role of particle size.

Few studies deal with the size distribution of manganese aerosols. Some measurements show that 80% of particles from a ferromanganese furnace have a size ranging from 0.1 to 1.0 μm. More than 99% of the particles in fumes around the blast furnace are smaller than 2 μm. On the other hand, manganese dust encountered in the mining of manganese ore usually contains only a small proportion of respirable particles (< 5 μm).

There is some evidence that aerosols formed by condensation may be more harmful than those formed by disintegration. Whether this is caused by the differences in the particle size distribution remains to be clarified. The toxicity of different manganese compounds appears also to depend on the type of manganese ion present and on the oxidation state of manganese. It seems that the less oxidized the compound the higher the toxicity.

In recent years many countries have established maximum allowable concentrations or threshold limit values (TLV) for manganese in the workroom air and working conditions have been improved. Therefore, the incidence of adverse health effects due to excessive exposure to manganese reported earlier may differ from the most recent data on manganese exposure. On the other hand, individuals differ markedly in their susceptibility to manganese. As a rule only a small percentage of those who are exposed to manganese aerosols develop symptoms and signs of chronic poisoning.

Biological indicators. As mentioned, the mean concentration of manganese in urine and faeces, and with less certainty, the mean concentration of manganese in blood, may give some information about the mean level of manganese exposure. Faecal collection and analysis for manganese content is not easy to perform. For this reason faecal manganese, although it could be used to monitor manganese exposure, has been measured only occasionally. On the other hand, determination of manganese in urine and blood is quite common. These tests have not yet been shown to be valid enough for the assessment of individual exposure. The present knowledge does not offer any other reliable biological measure that might be used as an indicator of individual manganese exposure.

4.2.2 *Analytical problems and their impact on interpretation of published data*

Precautions have to be taken against the contamination of samples when they are being taken and analysed. When air is being sampled, it is important to ensure that respirable particulate matter is collected and that the sample reflects the actual exposure of a worker or exposure at a working place. Inorganic solid samples are treated with acids to dissolve metallic oxides and to precipitate silica, which could interfere with the analysis of samples of manganese. Wet or dry ashing or fusion techniques with subsequent dilution of manganese in acid are usually adequate for preparing biological samples.

Owing to a very low manganese level in biological materials, it is necessary to concentrate the sample before analysis. The only exception is the measurement of manganese levels in blood with flameless atomic absorption spectrometry. The concentration of samples may cause matrix effects and it may therefore be necessary to separate the manganese from the sample.

The preparation of samples is the most delicate part of the procedure for manganese determination since the possibility of either loss or contamination increases with each additional step.

For the analysis of manganese, colorimetric methods are available.

The most sensitive method for the detection of manganese is atomic absorption spectroscopy. Methods for the estimation of manganese in urine and blood by means of flameless atomic absorption spectrometry have been described in the literature.

Direct spectrography, neutron-activation analysis, and a few other techniques (X-ray fluorescence, emission spectrometry) have been successfully used for the estimation of manganese concentrations in air and biological materials.

An intercomparison of analytical techniques carried out in 1974 by the European Atomic Energy Commission showed that there was a good agreement among the results obtained from the use of different techniques (neutron-activation analysis, X-ray fluorescence, emission spectrometry, and atomic absorption). Consequently, it seems that analytical problems, at least in recent years, should not have an impact on the interpretation of published data. However, contamination during sampling and analytical procedures may at least partly explain the rather wide range of manganese concentrations found in the literature particularly for levels in blood. For example, the use of heparin as an anticoagulant is still quite common although the manganese content in the heparin may exceed that in the blood sample.

The lack of information on the particle size distribution of manganese dust is probably one of the major problems in the interpretation of the results obtained in studies of the relationship between the level of manganese in air (at workplaces) and adverse health effects.

4.2.3 *Exposure-effect and exposure-response relationships*

Effects on the central nervous system. Manganism is reported in workers involved in the mining and processing of manganese ores and in the production of manganese alloys and dry cell batteries. As far as mining is concerned, the use of high-speed drilling machines, which produce a large amount of manganese dioxide dust, has been a major factor in the increased incidence of this disease.

In 1955, 150 cases of manganese poisoning were reported from Moroccan mines among a total of about 4000 workers. Of these, 132 occurred among miners working underground, drilling blast holes. The manganese concentration in air in the immediate vicinity of rock drilling was about 450 mg/m^3 in one mine and about 250 mg/m^3 in another (*24*). In a study of 72 Chilean miners exposed to concentrations of 62.5–250 mg/m^3, 12 (16.5%) were found to have manganese poisoning (*25*). A later study involving 370 miners exposed to manganese concentrations in air of 0.5–46 mg/m^3 showed that 15 workers (4%) developed manganism (*26*).

In two manganese ore crushing mills, 11 cases of manganese poisoning were found among 34 workers. In 9 workers exposed to concentrations below 30 mg/m^3 no case of manganism was detected, but 5 out of 8 men exposed for more than 3 years to concentrations exceeding 90 mg/m^3 had chronic manganese poisoning (*27*).

In a dry-cell battery plant, 8 out of 36 workers exposed to dust containing 65–70% manganese oxide with an average dust concentration in the range 4.8–28.4 mg/m^3 showed neurological and psychiatric changes. Six had chronic manganese psychosis, one had left hemiparkinsonism, and one had left choreoathetosis (*28*).

Studies were conducted in Japan in a manganese crushing and refining factory, in a dry-cell factory, and in a welding-rod factory (*29, 30*). In 4 out of 47 refinery workers exposed to manganese concentrations of 2.3–17.1 mg/m^3, disturbances of the central nervous system were clearly observed, and 11 were suspected of having some neurological signs. Similar findings were noted in 4 out of 32 workers employed in welding-rod manufacture who were exposed to manganese air levels of 3.1–8.1 mg/m^3. Seven out of 55 workers exposed to air manganese levels of 1.9–21.1 mg/m^3 at a dry-cell battery factory also showed some neurological disturbances. In this study a significant correlation was observed between the manganese content of urine samples and the neurological findings. In the manganese ore crushing and refining group, significant correlations were observed between the manganese content of whole blood and urine samples and neurological symptoms and signs.

An investigation carried out in two ferromanganese factories (*31, 32*) showed that in one factory 27 of 160 workers exposed to air concentrations ranging from 1.9 to 4.9 mg/m^3 in the sintering area of the plant and to 0.06–2.0 mg/m^3 in other areas showed some adverse effects. More than 30% complained of failing memory, fatigue, increased perspiration, and hyposexuality. Trembling fingers were found in 24%, 3.5% displayed muscle rigidity, 5% had adiadochokinesis, and 13% showed balance disturbance. The geometric mean manganese concentrations in blood and urine of 144 males out of a total of 160 subjects were 184 μg/l and 46 μg/l respectively. In the other factory, after exposure to manganese concentrations in air ranging between 0.6 and 8.6 mg/m^3 (occasionally values of about 24 mg/m^3 were recorded under the conveyor-belt), 40 out of 100 electric furnace workers complained of increased perspiration, failing memory, headache, and sleepiness. Eight of the workers had adiadochokinesis, 10 had trembling fingers, and 8 showed acceleration of the patellar reflex. The geometric mean of manganese in blood was 110 μg/l and in urine 45 μg/l.

According to another report involving 155 workers from 75 ferromanganese plants, 7 workers showed urinary concentrations of manganese above 8 μg/l and appeared to have manifestations of neurological disturbances, after exposure to airborne manganese concentrations above 5 mg/m^3 (*1*); 15 workers were suspected of having slight abnormalities.

A rather high prevalence of neurological symptoms and signs (in 167 out of 994 workers) was reported in a ferromanganese foundry with levels of exposure to manganese ranging between 0.45 and 0.6 mg/m^3 (*11*). In another ferromanganese foundry neurological effects were discovered in 91 out of 200 workers at exposure levels of 0.6–1.2 mg/m^3 (*33, 34*).

In a study (*35*) on 369 workers in a factory for the production of ferromanganese alloys, where the exposure was in the range 0.3–20.4 mg/m^3, two other groups of workers were used as controls. The first group consisted of 190 electrode plant workers who did not have any occupational exposure from their own work but were subject to low exposure from the ferromanganese plant on the same premises. The second group comprised 204 workers from an alluminium rolling mill situated about 5 km from the ferromanganese plant. At the time of examination 62 of the 369 in the study group had some neurological signs. Most workers (47) had only finger tremor, but 11 had pathological reflexes, and 4 had cogwheel phenomenon as an isolated finding or combined with tremor or pathological reflexes.

In the first control group whose exposure ranged between 2 and 30 μg/m^3, 10 of the 190 workers had finger tremor and one had pathological reflexes. In the second control group—exposure ranging between 0.05 and 0.07 μg/m^3—neither finger tremor nor any pathological reflex was observed.

However, the question arises whether the reported slight neurological signs in a small percentage of workers from the electrode plant can be attributed to manganese exposure considering that in 66 workers from the ferromanganese plant with an exposure between 0.47 and 1.06 mg/m^3 of air no neurological signs were found. It was also noticed that some of the recorded subjective symptoms, which might be the symptoms of the subclinical stage of manganism, in the ferromanganese group were more frequent in moderate and heavy smokers. Smokers from the ferromanganese factory showed some of the subjective symptoms more often than the smokers from the other two comparison groups. The mean concentration of manganese in the urine of ferromanganese workers was in the range of 1.3–21.7 μg/l.

Individual cases of manganese poisoning were also reported from a ferromanganese plant in which the concentrations of manganese in the air were in the range 0.1–4.7 mg/m^3 (*36*). There is a possibility that the reported concentrations of manganese were less than those actually present because a later report (*1*) indicated the average concentration of manganese in some areas of the plant in question to be 11 mg/m^3 of air. In another ferromanganese plant where 5 cases of manganism were found among 142 workers studied, one worker developed chronic manganese poisoning at an exposure level of only 0.33 mg/m^3 manganese estimated as weighted average (*37*).

It is noteworthy that in experiments on monkeys characteristic central nervous system signs were produced at rather low exposure levels. Thus, in one study, the exposure of a monkey to aerosols of manganese dioxide at concentrations of 0.6–3.0 mg/m^3 for 1 hour a day over a 4-month period caused symptoms and signs of extrapyramidal involvement (*38*). In another, more recent study (*39*), monkeys given subcutaneous injections of a suspension of manganese dioxide once a week for 9 weeks at doses of 0.25–1.0 g, developed typical extrapyramidal signs that appeared after 3–4 weeks. The severity of signs was not proportional to the dose administered but the time of their appearance was dose-related.

Table 7 summarizes the epidemiological studies pertinent to the evaluation of exposure-effect and exposure-response relationship (CNS involvement) in workers occupationally exposed to manganese.

Effects on the lungs. The association between exposure to manganese and a high rate of pneumonia was first noticed in men working in a pyrolusite mill, then among dry-battery workers, and then in workers handling pyrolusite. Although data on exposure levels for these cases are not available, it can be assumed that they were high.

An increased incidence of bronchitis, pharyngitis, and pneumomia was also registered in workers involved in the manufacture of potassium permanganate. The exposure conditions to dust were as follows: the manganese content of dust was between 41% and 66%; practically all particles were below 1 μm, 80% being below 0.2 μm; and manganese concentrations in air, calculated from the manganese dioxide content of dust, were between 0.1 and 13.7 mg/m^3. During the period 1938–45 the incidence of pneumonia in workers exposed to manganese averaged 26 per 1000 compared with an average of 0.73 per 1000 in a control group (*40*). Increased incidence of pneumonia was also described later in miners working in Moroccan manganese mines (*24*).

A higher rate of pneumonia was also described in workers exposed to the concentrations of manganese encountered in ferro-

Table 7. Epidemiological studies pertinent to the evaluation of exposure–effect and exposure-response relationships of the central nervous system

Authors	Number of workers and place of work	Manganese concentration in air	Observation
Flinn et al. (*27*)	34—manganese ore crushing mill	10.4–173 mg/m^3	11 developed manganism; 5 out of 8 workers, exposed to $>$90 mg/m^3 developed manganism after 3 years, and no case of manganism in exposure $<$30 mg/m^3
Ansola et al. (*25*)	72—manganese mine	62.5–250 mg/m^3	12 (16.5%) developed manganism
Rodier (*24*)	4000—manganese mine	250–450 mg/m^3	150 (37.5%) developed manganism
Schuler et al. (*26*)	370—manganese mine	0.5–46 mg/m^3	15 (4%) developed manganism
Whitlock et al. (*36*)	manganese steel plant	0.1–4.7 mg/m^3	2 cases of manganism
Gorodnova (*11*)	994—ferromanganese foundry	0.45–0.60 mg/m^3	167 (16.8%) workers with neurological signs and symptoms
Tanaka & Lieben (*1*)	155—75 ferromanganese plants	$<$5 mg/m^3 (38 workers) $\geqslant$5 mg/m^3 (117 workers)	no neurological signs; 7 (6%) developed manganism and 15 (12.8%) had suspected neurological signs
Horiguchi et al. (*29*); Horiuchi et al. (*30*)	47—manganese ore crushing and refining	2.3–17.1 mg/m^3	4 (8.5%) developed manganism; 11 (23.4%) with slight neurological abnormalities
	32—welding-rod manufacturing plant	3.1–8.1 mg/m^3	4 (12.5%) with slight neurological abnormalities
	55—dry-cell battery manufacturing plant	1.9–21.1 mg/m^3	7 (12.7%) with slight neurological abnormalities
Emara et al. (*28*)	36—dry-cell battery factory	4.8–28.4 mg/m^3	8 (22.2%) with neuropsychiatric manifestations
Smyth et al. (37)	142—ferromanganese plant	fumes: 0.12–13.3 mg/m^3 dusts: 2.1–12.9 mg/m^3	5 (3.5%) developed manganism; (3 exposed to highest concentrations of fumes and 2 exposed to highest concentrations of dusts) 1 case of manganism found when the time-weighted average exposure was 0.33 mg/m^3 for 10.5 years

Table 7 (continued)

Authors	Number of workers and place of work	Manganese concentration in air	Observation
Suzuki et al. (*31, 32*)	160—ferromanganese plant	0.06–4.9 mg/m^3	27 (16.9%) with slight neurological signs and symptoms
	100—ferromanganese plant	0.6–8.6 mg/m^3	40 (40%) with slight neurological signs and symptoms
Karnauch et al. (*34*) and Kovaltchuck & Brodski (*33*)	200—ferromanganese foundry	0.61–1.2 mg/m^3	91 (45.5%) workers with neurological signs and symptoms
Šarić et al. (*35*)	369—ferromanganese plant	0.3–20.4 mg/m^3	62 (16.8%) with slight neurological signs and one case of manganism
	190—electrode plant (adjoining)	0.002–0.03 mg/m^3	11 (5.8%) with slight neurological signs (finger tremor and in 1 case pathological reflexes)
	204—aluminium processing	0.05–0.07 μg/m^3	no neurological signs

manganese plants. In a retrospective study (*4*) a higher rate of sick-leave due to pneumonia was reported among workers involved in the production of manganese alloys with manganese exposure ranging from 0.15 to 20.4 mg/m^3. During the same retrospective study, a higher rate of bronchitis (acute, not specifically defined, and chronic) was found in workers occupationally exposed to manganese compared with two control groups.

However, there are some indications that even low exposure to only 1 μg/m^3 (environmental exposure) may be connected with an increase in the rate of acute bronchitis and pneumonia in the general population (*41*). In this particular study the pneumonia incidence as a whole did not seem to exceed the predicted figures but it did not differ in relation to the seasons either. The fact that manganese concentrations were usually higher in summer than in winter suggests that this may have contributed to the uniform number of pneumonia cases in summer and winter. Even if the incidence of the disease is associated with exposure, it is still a moot point whether the respiratory effect can be exclusively attributed to manganese since it is possible that it is due to the action of manganese aerosol and sulfuric acid, i.e., sulfates absorbed on the surface of manganese particles. Furthermore, it is possible that some other factors that were not sufficiently controlled might have influenced the results.

There are no data on the basis of which calculations can be made regarding exposure-response relationship for pneumonia caused by occupational exposure to manganese in humans.

4.2.6 *Conclusions*

The available data suggest that in assessing the health impact of manganese exposure the following effects should be considered as critical:

— effects on the central nervous system with signs of extrapyramidal involvement, and

— effects on the lungs causing an increased incidence of pneumonia.

Owing to the wide ranges of manganese concentrations in the workroom air reported in most of the studies reviewed and to the lack of more precise information, it is difficult to judge whether exposure limits should be based on central nervous system effects or on lung effects.

On the basis of the data presented on exposure–effect and exposure–response relationships it appears that adverse effects on the central nervous system occur in some workers at manganese levels of 2–5 mg/m^3 of air. Experimental studies suggest that such effects may be induced at concentrations around 1 mg/m^3. There are indications that symptoms and signs, which are not specific but may be connected with the early stage of manganism, can be found in a number of workers who have been exposed to manganese concentrations of about 0.5 mg/m^3 of air; at approximately the same exposure levels certain susceptible individuals may develop chronic manganese poisoning. Adverse effects on the lungs of manganese-exposed workers do not seem to appear at concentrations below 0.3–0.5 mg/m^3 of air.

It has to be pointed out that the increased incidence of pulmonary disease found at exposures to the rather low concentrations of manganese reported in some studies is not necessarily attributable directly to the manganese itself. The contention is that manganese exposure increases susceptibility to the occurrence of pneumonia or other acute respiratory diseases by disturbing the normal mechanism of lung clearance. A disturbance of the clearance mechanisms can also increase the amount of manganese retained in the lung, and particles generally not considered respirable may also be deposited.

4.2.7 *Problems of deriving health-based occupational exposure limits for populations with a high non-occupational exposure*

Drinking-water in the neighbourhood of a manganese mining area may contain a high concentration of manganese. This source of exposure as well as exposure from food contamination ought to be considered. It has been observed that entry of manganese via the gastrointestinal tract can significantly contribute to intoxication in miners.

4.3 Research possibilities

Mechanisms involved in the uptake and clearance of manganese from the respiratory system and the gastrointestinal tract are still almost completely unknown. The same applies to the factors that affect these processes.

Concentrations of manganese in the tissues at which adverse health effects are observed should be established.

Additional studies are required to determine the size of airborne manganese particles, so that total intake through the respiratory pathway can be estimated more precisely.

Additional studies are required on the size distribution of airborne manganese in various industrial processes. Further studies are also needed on the difference in biological activity between aerosols formed by condensation and those formed by disintegration.

The individual differences that have been observed in human susceptibility to exposure to manganese have not been explained. The question also arises whether there are group differences in susceptibility.

It is still unknown what accounts for the time course of the symptoms and signs of manganism. Little is known about the significance of the appearance of symptoms and signs of central nervous system involvement, which can be found in a proportion of workers at rather low exposure levels, in the development of more advanced stages of manganism.

More research is needed on the relationship between manganese concentrations in air and urinary and blood levels of manganese as well as on the relationship between the latter biological measures and early neurological signs and symptoms.

Knowledge of the exposure–response characteristics of manganese in relation to the pulmonary and central nervous system effects is also lacking. The same applies to the interactions of manganese with other pollutants, diet, age, and the general health status of the exposed workers.

From animal experimental data it appears that disturbances in sexual function and germinal activity occur as a result of exposure to manganese; these disturbances should be investigated in humans.

Further studies are also required on the potential carcinogenicity, teratogenicity, and mutagenicity of manganese. The effect, if any, of manganese on the myocardium also needs to be evaluated.

4.4 Recommendations

On the basis of some of the quoted studies and observations in ferromanganese plants, where workers are exposed to dusts and fumes containing mainly manganese dioxide in particles small enough to reach the alveoli, a value of 0.3 mg of respirable manganese particles per m^3 of air (time-weighted average) can be recommended as the health-based limit of occupational exposure.

However, the proposed value is *tentative* in view of the fact that there is a lack of precise information about exposure–effect and exposure–response relationships.

Since in some instances non-occupational exposure to manganese may be high, the authorities must take this additional exposure into consideration in setting exposure limits.

The potential biological indicators of exposure to manganese have not yet been sufficiently validated for monitoring individual exposure; therefore assessment of the exposure of workers to manganese still has to rely on the determination of manganese air levels.

REFERENCES

1. TANAKA, S. & LIEBEN, J. *Archives of environmental health*, **19**: 674 (1969).
2. TSALEV, D. L. ET AL. *Bulletin of environmental contamination toxicology*, **17**: 660 (1977).
3. JINDŘICHOVA, J. *Internationales Archiv für Gewerbepathologie und Gewerbehygiene*, **25**: 347 (1969) (in German).
4. FOLPRECHTOVA, A. ET AL. *Pracovní Lékarství*, **22(3)**: 92 (1970) (in Czech).
5. ROSENSTOCK, H. A. ET AL. *Journal of the American Medical Association*, **217**: 1354 (1971).
6. COTZIAS, G. C. ET AL. *Nature*, **201**: 1228 (1964).
7. ZIELHUIS, R. L. ET AL. *Environmental health perspectives*, **25**: 103 (1978).
8. BUCHET, J. P. ET AL. *Clinica chimica acta*, **73**: 481 (1976).
9. SUZUKI, Y. ET AL. *Tokushima journal of experimental medicine*, **22**: 5 (1975) (in Japanese).
10. EADS, E. A. & LAMBDIN, E. C. *Environmental research*, **6**: 247, (1973).
11. GORODNOVA, N. B. [*Clinical picture and cause of manganese poisoning at simultaneous exposure to aerosols of manganese oxides formed by condensation, local vibrations and physical strain*], Sverdlovsk, Institut Gigieny Truda i Professionalnyh Zabolevanij (1967) (in Russian).
12. NEIZVJESTNOVA, E. M. [The toxicity of manganese chloride under conditions of general vibration]. In: [*Combined effects of chemical and physical factors of the working environment*], Sverdlovsk, Institut Gigieny Truda i Professionalnyh Zabolevanij, 1972, p. 76 (in Russian).
13. BERGSTRÖM, R. *Scandinavian journal of work, environment and health*, **3**: Suppl. 1 (1977).
14. ŠARIĆ, M. & LUČIĆ-PALAIĆ, S. Possible synergism of exposure to airborne manganese and smoking habit in occurrence of respiratory symptoms. In: *Inhaled particles IV*, Oxford and New York, Pergamon Press, 1977.
15. SAAKADZE, P. B. & VASILEV, B. G. *Gigieny truda i professional'nye zabolevanija*, **6**: 19 (1977) (in Russian).
16. ŠARIĆ, M. & HRUSTIĆ, O. *Environmental research*, **10**: 314 (1974).
17. MANDŽGALADZE, R. N. *Voprosy gigieny truda i profpatologii*, **11**: 126 (1967) (in Russian).
18. MANDŽGALADZE, R. N. & VASAKIDZE, I. M. *Voprosy gigieny truda i profpatologii*, **10**: 209 (1966) (in Russian).
19. SIROVER, M. A. & LOEB, A. L. *Science*, **194**: 1434 (1976).
20. LOEB, L. A. ET AL. *Journal of toxicology and environmental health*, **2**: 1297 (1977).
21. HOFFMAN, D. J. & NIYOGI, K. S. *Science*, **198**: 513 (1977).

22. WEYMOUTH, L. A. & LOEB, L. A. *Proceedings of the National Academy of Sciences of the United States of America*, **75**: 1924 (1978).

23. STONER, G. D. ET AL. *Cancer research*, **36**: 1744 (1976).

24. RODIER, J. *British journal of industrial medicine*, **12**: 21 (1955).

25. ANSOLA, J. ET AL. *Revista medica de Chile*, **72**: 222 (1944) (in Spanish).

26. SCHULER, P. ET AL. *Industrial medicine and surgery*, **26**: 167 (1957).

27. FLINN, R. M. ET AL. *Chronic manganese poisoning in an ore-crushing mill*, Washington, DC, US Government Printing Office (Public Health Bulletin No. 247), 1940.

28. EMARA, M. A. ET AL. *British journal of industrial medicine*, **28**: 78 (1971).

29. HORIGUCHI, S. ET AL. *Japanese journal of industrial health*, **8(6)**: 333 (1966) (in Japanese).

30. HORIUCHI, K. ET AL. *Osaka City medical journal*, **16(1)**: 29 (1970).

31. SUZUKI, Y. ET AL. *Shikoku acta medica*, **29(6)**: 425 (1973) (in Japanese).

32. SUZUKI, Y. ET AL. *Shikoku acta medica*, **29(6)**: 433 (1973) (in Japanese).

33. KOVALČUK, A. A. & BRODSKI, B. O. [Health of the workers in ferroalloys foundries]. In: [*Proceeding of a Conference on Occupational Hygiene and Prevention of Occupational Diseases in Ferroalloys Production*], Sverdlovsk, Institut Gigieny Truda i Professionalnykh Zabolevanij, 1973, p. 78 (in Russian).

34. KARNAUCH, N. G. ET AL. [Air pollution in old and new foundries of ferroalloys]. In: [*Proceedings of a Conference on Occupational Hygiene and Prevention of Occupational Diseases in Ferroalloys Production*], Sverdlovsk, Institut Gigieny Truda i Professionalnyh Zabolevanij, 1973, p. 27 (in Russian).

35. ŠARIĆ, M. ET AL. *British journal of industrial medicine*, **34**: 114 (1977).

36. WHITLOCK, C. M. ET AL. *American Industrial Hygiene Association journal*, **27**: 454 (1966).

37. SMYTH, L. T. ET AL. *Journal of occupational medicine*, **15**: 101 (1973).

38. VON BOGAERT, L. & DALLEMAGNE, M. J. *Monatsschrift für Psychiatrie und Neurologie*, **111**: 60 (1945–46) (in German).

39. SUZUKI, Y. ET AL. *Tokushima journal of experimental medicine*, **22**: 5 (1975) (in Japanese).

40. LLOYD-DAVIES, T. A. *British journal of industrial medicine*, **3**: 111 (1946).

41. ŠARIĆ, M. *Biological effects of manganese*. Research Triangle Park, NC, US Evironmental Protection Agency, 1978, p. 152 (EPA-600/1–78–001).

42. ŠARIĆ, M. ET AL. Acute respiratory diseases in a manganese contaminated area. In: *Proceedings of the International Conference on Heavy Metals in the Environment. Toronto, 1975, Volume III*, Toronto, Institute for Environmental Studies, University of Toronto, 1978, p. 389.

5. INORGANIC MERCURY[1]

5.1 Introduction

Mercury exists in a wide variety of physical states, all of which have intrinsic chemical properties and require independent toxicological assessment. This section deals only with elemental mercury and its inorganic compounds, with special emphasis on the former. The latter constitutes the most common occupational exposure and there is substantial information on it. Mixed exposure—i.e., mercury vapour and mercury(II)—is not uncommon in industry though it is not often considered in the literature. It must be mentioned that inorganic mercury when released into the environment may form methylmercury, which may accumulate in aquatic fauna and consequently a high fish diet may contribute to the mercury body burden. The most hazardous forms of mercury to human health are elemental mercury vapour and the short-chain alkylmercurials.

5.2 Metabolism

5.2.1 *Absorption, distribution, and excretion*

Approximately 80% of inhaled mercury vapour is initially retained. Information on pulmonary retention of other forms of mercury in man is lacking. In general, aerosols of mercury follow the same general physical laws governing deposition in the respiratory system as other aerosols. Particles deposited in the upper respiratory tract are, as a rule, cleared quickly but those deposited in the lower respiratory tract are retained and mercury may be absorbed from them depending on the solubility of the compound.

Liquid metallic mercury is poorly absorbed from the gastrointestinal tract—probably less than 0.01% of the ingested dose. Absorption of inorganic mercury(II) compounds from food is about 7% of the ingested dose. Little information is available on absorption through the skin although it is suspected that most forms of mercury can penetrate the skin to some extent.

Animal data indicate that the kidneys accumulate the highest tissue concentrations no matter what form of mercury is admin-

[1] This section is mainly based on publications *1, 2* and *3* in the reference list.

istered. Inhaled elemental mercury vapour is distinguished from other inorganic mercury compounds by its ability to cross the blood–brain barrier and placenta rapidly. Information on placental transfer is based on animal data only. Metallic mercury is rapidly transformed to divalent ionic mercury in the body. In the WHO publication *Environmental Health Criteria 1. Mercury* (*1*; p. 75) it is stated:

> "Despite the rapid oxidation that has been shown to take place in the red blood cells, some elemental mercury remains dissolved in the blood long enough for it to be carried to the blood–brain barrier and to the placenta. Its lipid solubility and high diffusibility allow rapid transit across these barriers. Tissue oxidation of the mercury vapour in brain and fetal tissues converts it to the ionic form which is much less likely to cross the blood–brain and placental barriers. Thus oxidation in these tissues serves as a trap to hold the mercury and leads to accumulation in the brain and fetal tissues."

The distribution of mercury between erythrocytes and plasma depends upon the form of the mercury. The erythrocyte to plasma ratio is approximately 1 for inorganic mercury compared to approximately 10 for methylmercury (*4*).

Most forms of mercury are eliminated mainly with the urine and faeces. In long-term exposure to mercury vapour, urinary excretion of mercury slightly exceeds faecal elimination. Large individual fluctuations in daily mercury excretion are common even when exposure conditions are the same (*1, 5, 6*). There are no data to show that even in microenvironmental exposure (*7*) urinary levels of mercury are good indicators of individual exposure; furthermore, urinary levels are completely unsatisfactory for methylmercury exposure since 90% of it is eliminated with faeces.

The biological half-lives of mercury compounds other than methylmercury are not well established and this is particularly true for the organs that are of toxicological importance. Little information exists on the biological half-life of mercury after exposure to mercury vapour. In 5 volunteers who inhaled mercury vapour for approximately 15 minutes, the elimination of mercury from the body followed a single exponential process and a biological half-life of 58 days (the range of individual values being 35–90 days) was recorded (*8*). The half-life of mercury elimination from the whole body may not be a reliable indicator of mercury accumulation in specific organs because it seems that the rate of mercury elimination varies from organ to organ. It has been reported (*1, 9, 10*) that the concentration of mercury in the brain of an individual was high even 10 years after the cessation of exposure, suggesting that the brain does not follow the same kinetics of elimination as the whole body.

5.2.2 *Concentration of mercury in biological materials as an indicator of exposure*

According to the available information the mercury body burden of an individual exposed to a constant average concentration of mercury vapour does not reach a steady state until after about one year of exposure. Consequently the levels of mercury in blood and urine cannot be expected to reach steady levels until the individual has had one year of exposure.

Experience from occupational health studies shows that on a group basis there is an approximately linear relationship between the time-weighted average air concentration and urinary concentration of mercury. A similar relation also exists between concentration of mercury in urine and concentration in blood.

According to *Environmental Health Criteria 1. Mercury* (*1*), a time-weighted average air concentration of 50 μg of mercury per m^3 would approximately correspond to 35 μg of mercury per litre of blood and 150 μg per litre of urine. These conclusions were based on the work done by Smith et al. (*11*). The original data in their study, however, indicate that the ratio between the concentration in air and the concentration in urine is 1 : 2–2.5 and not 1 : 3 as concluded in the WHO publication (*1*). A recent study (*12*) suggests that this ratio is close to 1 : 1 but another (*13*) puts it at 1 : 2. The Study Group considered the latter as reliable; the exposure limits recommended in this document are based on that ratio.

There is no evidence that concentrations of mercury in blood and urine can be used to evaluate exposure to forms of mercury other than metallic mercury vapour.

5.2.3 *Concentrations of mercury in biological materials as indicators of body burden and concentrations in critical organs*

A complete metabolic model in man following exposure to elemental mercury vapour or inorganic mercury salts is not available and at present it is not possible to relate concentrations of mercury in indicator materials to mercury body burden or to concentrations in the critical organs such as the central nervous system after inhalation of elemental mercury (*1*).

5.2.4 *Mercury levels in blood, hair, and urine in non-occupationally exposed subjects*

Mercury levels in blood may be highly influenced by fish intake. People who do not usually eat fish and have no occupational exposure have mercury levels in whole blood of below or about 5 μg/l. Moderate consumption of fish containing mercury may give blood levels of 10–20 μg/l. In heavy fish eaters values of 100–200 μg/l or even more may be observed.

Mercury levels in hair are also to a great extent influenced by fish consumption. When the consumption of contaminated fish is not high, values of less than a few milligrams per kilogram are found, while with heavy consumption of contaminated fish values of about 20–50 mg/kg or higher may be observed.

The general level of mercury in urine in non-exposed subjects seems usually to be below or about 0.5 μg/l, although much higher values have also been reported. Since mercury in fish is in the form of methylmercury, which contributes little to mercury in urine, there is little need to take fish intake habits into account when measuring mercury in urine (*1, 2*).

5.3 Toxic effects

5.3.1 *Acute effects*

Severe intoxication by inorganic mercury can be provoked by:

(1) accidental short-term inhalation of high concentrations of elemental mercury vapour, causing bronchial irritation, erosive bronchitis, and diffuse interstitial pneumonitis, and

(2) ingestion of electrolytic inorganic salts of mercury(II) that can produce local necrotic changes in the gastrointestinal tract, circulatory collapse, or acute renal failure with oliguria or anuria.

5.3.2 *Chronic effects*

Early stages of chronic poisoning by inorganic mercury, usually by industrial exposure to elemental mercury vapour alone or in combination with dust of mercuric salts, are characterized by anorexia, loss of weight, and minor symptoms of central nervous system dysfunction. The symptoms include increased irritability, loss of memory, loss of self-confidence, and insomnia. Later phases are characterized by mercurial tremor, psychic disturbances, and changes in personality (*1, 6, 12, 14*), although there are reports where slight tremor has been considered as an early sign (*15*).

Symptoms of chronic mercury exposure include inflammation of the gums with swollen and bleeding margins. These signs are difficult to distinguish from signs of pyorrhoea and, moreover, these observations date back to the time when industrial exposure was often high and standards of hygiene were often low. These changes cannot thus be considered in deriving health-based occupational exposure limits.

In exceptional cases, chronic occupational exposure may produce proteinuria and symptoms of nephrotic syndrome (*1, 6, 16*). Recent data indicate that subclinical effects on the kidneys may be more frequent than has been recognized hitherto (*17*). Deposition of mercury on the anterior surface of the eye lens (mercuria lentis) is only a sign of exposure, not a symptom of chronic mercurialism (*1*).

Idiosyncrasy as a result of exposure to trace amounts of inorganic mercury is reported in older literature, mainly in connection with local application of mercury preparations. A specific form of systemic reaction to mercury—acrodynia (pink disease or erythredema polyneuropathy)—has been described in children. There is no doubt that this systemic reaction was to some extent related to exposure to mercury, but because the disease was almost eradicated before a complete analysis of the mechanisms inducing it could take place, the relationship to mercury exposure has never been definitely established (*1*).

5.3.3 *Carcinogenic and genotoxic effects*

There exists no evidence that inorganic mercury is carcinogenic. In some experimental studies results showing chromosome breaks and mitotic aberrations after mercury exposure have been reported (*1*).

5.3.4 *Effects relevant to recommended health-based occupational exposure limits*

In the case of high exposure, the effects on the respiratory organs after inhalation of elemental mercury vapour and on the gastrointestinal tract after ingestion of inorganic mercury(II) may be considered as relevant. However, in chronic poisoning, the effects that are particularly relevant are the effects on the central nervous system arising from the inhalation of mercury vapour and the effects on the kidneys resulting from exposure to mercury(II).

5.4 Relationship between exposure and health effects

5.4.1 *Environmental and biological indicators of exposure*

No ideal biological indicator is available for evaluating the risk of mercury intoxication through the inhalation of mercury vapour. For the assessment of individual exposure neither mercury in blood nor mercury in urine is satisfactory as an indicator. However, on a group basis mercury levels in blood and urine parallel exposure, but the levels are influenced by recent exposure and reach a steady state with exposure only after continuous exposure for one year or more. Owing to a longer biological half-life of mercury in the brain than in the body as a whole, discontinuation of exposure may result in low urinary levels although high concentrations continue in the brain; furthermore, there are great individual variations (*1*). High mercury levels in urine may serve as a warning even in individual cases, however. Research is being conducted in the chlorine industry in the USA (*15*) and elsewhere (*7, 18*) on a more extensive use of mercury concentrations in urine for evaluating exposure and risks.

5.4.2 *Analytical problems*

Not only are there errors in the determination of mercury, but significant and often major errors occur during the collection, transport, and storage of the samples. These shortcomings should be borne in mind when considering the accuracy of determination of mercury in air and biological samples.

There are different ways of collecting samples for the determination of mercury in air. One method for the determination of total mercury consists in using two bubblers in series, containing sulfuric acid and potassium permanganate. The airborne mercury trapped in the two bubblers is determined by the atomic absorption procedure, which has a collection efficiency for total mercury of more than 90% when the mercury is in the form of elemental vapour or inorganic salts.

Commercially available portable devices can be used to determine mercury levels in air. The air is pumped through an optical cell that measures the absorption of light emitted from a mercury vapour lamp. These units, although convenient, measure only elemental mercury vapour and are subject to interference from other substances present in the working environment. They should be calibrated each time before use. These units also suffer from the deficiency of sampling only small volumes of air, which may not give a representative picture of the working environment. Research should be directed towards the development of personal monitoring devices.

The methods of collecting and storing samples of blood, hair, and urine are of great importance. With respect to blood samples, care should be taken to avoid any clot formation. Evacuated, heparinized specimen tubes are convenient for blood collection. It should be borne in mind that some commercially available anticoagulants may be contaminated with considerable amounts of mercury.

Measurements of the very low levels of mercury found in the uncontaminated general environment make special demands on the skills of the analyst. A method for the determination of mercury in nanogram quantities cannot be regarded as a routine procedure. Where conditions allow, it is highly desirable that the results obtained with one method be checked against those obtained by a different method and that the measurements made in one laboratory be compared with those made in another.

Determination of mercury by colorimetric measurement of a mercury dithizone complex was the basis of most of the methods used in the 1950s and the 1960s. These methods made use of wet oxidation of the sample, followed by extraction of mercury in an organic solvent as a dithizone complex and finally the colorimetric determination of the complex itself. The dithizone procedure has an absolute sensitivity of about 0.5 μg of mercury.

Today, the most widely used method for determining total mercury in environmental and biological samples is the flameless atomic absorption method which is rapid, sensitive and technically simple. Neutron activation is used as a reference method against which the accuracy of atomic absorption procedures may be checked (*1*).

Methods based on the formation of coloured compounds of mercury with copper and iodine were widely used in the 1960s in the USSR for analysis of mercury in air and urine. The precision and accuracy of these methods are considered to be low (*6*).

5.4.3 *Definition of critical organ*

The critical organ after short-term exposure to high concentrations of mercury vapour is the lung, where symptoms of pulmonary irritation will appear. After similar exposure to mercuric salts the kidney is the critical organ (*1, 6*).

In long-term exposure, the central nervous system is the critical organ for the toxic effects of inhaled elemental mercury. The risk to human health from long-term exposure to inorganic mercury other than metallic mercury is difficult to estimate because there are no recorded cases of human poisoning. Judging from animal data the critical organ for mercury(II) should be the kidney (*1, 6*).

5.4.4 *Exposure-response relationships for critical adverse effects*

Acute pulmonary effects have been observed after exposure to very high concentrations of mercury in air. Although the exact concentration that gives rise to such effects is not known, it is estimated that exposure to concentrations of 1–3 mg of mercury per m^3 of air can cause acute pneumonitis (*6*).

There is very little information on human response to poisoning with mercury salts. Most reported cases involve people who took mercury(II) chloride with the intention of committing suicide. It appears that severe poisoning occurs after ingestion of less than 1 g of mercury(II) chloride.

In the literature there is some information on the mercury content of organs after fatal mercury(II) poisoning. Mercury concentrations in the kidneys have ranged from 9 to 70 mg/kg wet weight and in the liver from 3 to 63 mg/kg wet weight (*6*).

Exposure-response relationships are not known for most exposure situations. However, some data are available for metallic mercury vapour exposure that make the evaluation of exposure risks possible to some extent. Most data on long-term exposure to mercury vapour come from industry and from animal experiments.

Prolonged exposure in an industrial environment to about 100 μg of mercury per m^3 of air involves the risk of mercury intoxication with classical tremor symptoms (*1, 6, 11*). Data from studies carried out in the USA and the USSR show that some effects appear at lower concentrations.

An increase in complaints of loss of appetite and insomnia was observed in a group of US workers exposed to time-weighted average air concentrations of mercury of 60–100 $\mu g/m^3$ when compared to two other groups subjects to a lower exposure range of 10–50 $\mu g/m^3$ (*9*). Some effects were observed even at concentrations below 50 $\mu g/m^3$ (*6*). Reports from the USSR suggest that exposure to as little as 10–50 $\mu g/m^3$ gives rise to signs of micromercurialism (*14*).

However, it appears that the occurrence of adverse effects below 0.05 mg/m^3 has not been unequivocally established (*1*).

Animal experiments suggest that exposure to a few milligrams of mercury vapour per cubic metre of air for a few months (8 h daily) gives rise to lethal effects. Pathological changes in the brain and kidneys have been reported to occur at exposure levels of about 1 mg/m^3. Enzymatic changes in the blood, heart, liver, and kidneys have also been reported (*1, 6, 19*).

Studies in the Russian literature report signs of weight loss and effects in several animal species at very low exposure levels. After exposure to mercury at 10–30 μg/m^3 of air for several months, changes in the function of several organs of rats and rabbits were observed. These included: the central nervous system (conditioned reflexes), thyroid (increased uptake of iodine-131), heart (changes in ECG), liver (changes in the thymol turbidity test, protein synthesis, and ascorbic acid content), and adrenal gland (diminished ascorbic acid content and a slight weight increase). Disturbances in the immunological response of the body were also seen. CNS changes in conditioned reflexes in rats have been reported (*6, 18*) even at concentrations of only 2–5 μg of mercury per m^3 of air (exposure lasting several months).

It has been concluded (*1*) from the available evidence that long-term exposure for 8 h per day for 225 working days in a year to 100–200 μg of mercury per m^3 of air as a time-weighted average would, in humans, give rise to signs of tremor, and similar exposure to about 50 μg/m^3 would be associated with nonspecific symptoms. This conclusion still seems to be reasonable even though some data (*15*) show that effects occur at higher concentrations and other data (*12*) show the contrary.

The risk to human health from long-term exposure to inorganic mercury compounds is difficult to assess because there are no records of cases of human poisoning under these circumstances. From animal studies, a number of effects on various organs have been recorded as resulting from long-term exposure to inorganic mercury. After long-term exposure from ingestion of mercury salts, damage to the kidneys has been observed at dose levels in the diet exceeding 30 mg/kg of body weight (*1, 6*).

5.4.5 *Conclusions*

Although for elemental mercury it is possible to make evaluations which, on a group basis, relate exposure to symptoms as well as to concentration in biological materials, sufficient information is not available to permit precise quantification of risks. However, a small proportion of workers may be expected to have tremor after long-term exposure to 100 μg/m^3 and nonspecific symptoms at 50 μg/m^3. The possible occurrence of micromercurialism at levels below 50 μg/m^3 cannot be ruled out.

The Group felt that a health-based occupational exposure limit of 25 μg/m^3 (time-weighted average) would ensure a reasonable degree of protection not only against tremor but also against mercury-induced nonspecific symptoms.

In the case of inorganic mercury compounds such a limit cannot be defined on the basis of either animal or epidemiological data. Although it is not possible to identify even approximate minimum-adverse-effect levels for inorganic mercury compounds, it can be concluded (*1*) that the limited experience from occupational exposure suggests that this form of mercury is probably less hazardous than elemental mercury vapour. Thus the figures for occupational exposure to elemental mercury vapour would serve as conservative occupational exposure limits for this form of mercury (*1*). In view of the lower toxicity of inorganic mercury in other forms than elemental mercury, a limit of 50 μg/m^3 (time-weighted average) can be recommended as the health-based occupational exposure limit for these compounds.

Elemental mercury may give rise to acute pulmonary effects; therefore an upper limit should be also included in the health-based occupational exposure limit. In view of the fact that there are no reports indicating the occurrence of acute respiratory effects at the levels of 100–200 μg/m^3—levels common in industry in the past 20 years—and of the possibility of effects appearing at values above 1000 μg/m^3, the health-based upper limit will be between 200 and 1000 μg/m^3 for short-term exposure. Unfortunately there are insufficient data to specify the precise duration of such exposure.

If populations are exposed to high background levels of inorganic mercury the recommended health-based occupational exposure limit should be lowered accordingly.

There are no direct data on which to recommend a health-based biological exposure limit on an individual basis. On a group basis 25 μg Hg/m^3 would correspond to approximately 50 μg Hg/l of urine and should therefore not be exceeded. It seems prudent, however, to use this figure also on an individual basis taking into consideration the possibility of additional exposure arising from lack of personal hygiene.

In principle a health-based biological level for mercury in blood (Hg–B) could be recommended but since Hg–B levels are strongly influenced by mercury intake in food, separate specifications for different mercury forms would have to be designated. Data from animal experiments show that elemental mercury readily passes the placental barrier. In the absence of evidence to the contrary it seems prudent to take this as an indication of possible fetotoxic effects. Therefore, the exposure of women of child-bearing age to mercury vapour should be as low as pssible. The Group was not in a position to recommend a specific value.

5.5 Research possibilities

There is an immediate need for more epidemiological studies on exposure-response relationships, particularly with regard to mercury vapour, and a special emphasis should be placed on subclinical effects. Differences in susceptibility among individuals and subgroups of the population are not known. Such differences may be important in estimating acceptable exposure levels.

Some data tend to indicate that exposure to elemental mercury may give rise to adverse effects at considerably lower concentrations than have been hitherto recognized. In view of this fact, some studies should be repeated in order to eliminate any potential analytical and epidemiological errors. One major drawback to the epidemiological studies carried out in industry to date is the lack of coordination among them. Although there are ample opportunities to carry out epidemiological studies, they are reported from very few quarters. However, this situation is not unique to mercury.

Although the central nervous system seems to be the critical organ in exposure to mercury vapour, it would be useful to study the occurrence of subclinical signs of kidney dysfunction. Some data may indicate that the kidney, under certain circumstances, is a critical organ.

Though valuable information concerning uptake, distribution, and excretion is available, many more data are needed. Despite the fact that exposure to metallic mercury has occurred for a very long time, the biological half-life and the risk of accumulation of mercury in human beings, particularly in the central nervous system, is not known in any detail. Until such knowledge is available, difficulties will always exist in establishing health-based exposure limits and safety margins will have to be substantial.

More data are needed for evaluating the possibilities of using mercury levels in biological materials as reliable biological indicators (particularly urine) to assess the accumulation of mercury in different organs and hence exposure risks. Though the effects of methylmercury on the fetus have been studied, relatively little is known about the effects of inorganic mercury on the fetus and on reproduction in general. Additional studies on the possible reproductive and fetotoxic effects of inorganic mercury are required.

5.6 Recommendations

The recommended health-based occupational exposure limits for inorganic mercury include limits for mercury in both air and urine. All these values have to be considered as complementary to each other and not supplementary.

5.6.1 *Recommended health-based occupational exposure limits for metallic mercury vapour*

(1) *Long-term exposure.* In order to prevent symptoms of the central nervous system both air and urinary values are recommended:

air: 25 μg/m³ (time-weighted average);

urine: 50 μg/g creatinine (individual levels).

It is essential not to rely on the air limit alone since it will not necessarily reflect the total exposure.

Owing to the potential fetotoxicity of mercury, the above limits are not applicable to women of child-bearing age, whose exposure should be kept as low as possible.

(2) *Short-term exposure.* In order to prevent potential acute pulmonary effects a health-based limit is necessary. This value should be in the range 200–1000 μg/m³. This wide range reflects the uncertainty of the available data. The Study Group, despite the very limited information, recommends a value of 500 μg/m³ provided the time-weighted average is not exceeded.

5.6.2 *Recommended health-based occupational exposure limits for inorganic mercury compounds*

There is reason to consider inorganic mercury as less toxic than elemental mercury. It is therefore reasonable to propose a health-based occupational exposure limit of 50 μg/m^3. This value should also prevent potential effects on the renal system.

A corresponding value for mercury in urine cannot be recommended owing to lack of necessary data.

REFERENCES

1. WORLD HEALTH ORGANIZATION, *Environmental health criteria 1. Mercury.* Geneva, 1976.
2. FRIBERG, L. & VOSTAL, J., ed. *Mercury in the environment: An epidemiological and toxicological appraisal.* Cleveland, Chemical Rubber Company Press, 1972.
3. NORDBERG, G. F. ED. *Effects and dose response relationships of toxic metals. Proceedings of an international meeting organized by the Subcommittee on the Toxicology of metals of the Permanent Commission and International Association on Occupational Health. Tokyo, 18–23 November 1974.* Amsterdam, Elsevier, 1976.
4. BAKIR, F. ET AL. *Science,* **181**: 230 (1973).
5. FRIBERG, L. *Pure and applied chemistry,* **3**: 289 (1961).
6. FRIBERG, L. & NORDBERG, G. F. Inorganic mercury—relation between exposure and effects. In: Friberg, L. & Vostal, J., ed. *Mercury in the environment: An epidemiological and toxicological appraisal.* Cleveland, Chemical Rubber Company Press, p. 113 (1972)
7. STOPFORD, W. ET AL. *American Industrial Hygiene Association journal,* **39**: 378 (1978).
8. HURSH, J. B. ET AL. *Archives of environmental health,* **31**: 302 (1976).
9. TAKAHATA, N. ET AL. *Folia psychiatrica et neurologica japonica,* **24**: 59 (1970).
10. WATANABE, S. Mercury in the body 10 years after long-term exposure to mercury. *Proceedings of the Sixteenth International Congress on Occupational Health, Tokyo, September 22–27* (1969).
11. SMITH, R. G. ET AL. *American Industrial Hygiene Association journal,* **31**: 687 (1970).
12. GAMBINI, G. *La medicina del lavoro,* **69**: 379 (1978).
13. LINDSTEDT, G. ET AL. *Scandinavian journal of work, environment and health,* **5**: 59 (1979).
14. TRACHTENBERG, I. M. [*The chronic action of mercury on the organism, current aspects of the problem of micromercurialism and its prophylaxis.*] Zdorov'e, Kiev, 1969 (Russian with German translation).

15. LANGOLF, G. D. ET AL. *American Industrial Hygiene Association journal*, **39**: 976 (1978).

16. KAZANTZIS, G. ET AL. *Quarterly journal of medicine*, **31**: 403 (1962).

17. BUCHET, J. P. ET AL. Relationship between exposure to heavy metals and prevalence of renal dysfunction. *Archives of toxicology* (in press).

18. CHERIAN, M. G. ET AL. *Archives of environmental health*, **33**: 109 (1978).

19. ASHE, W. F. ET AL. *A.M.A. archives of industrial hygiene and occupational medicine*, **7**: 19 (1953).

6. CONCLUSIONS AND RECOMMENDATIONS

1. The Study Group concludes that the WHO programme on health-based permissible levels for occupational exposure to toxic substances is an important and essential activity. On the basis of current knowledge, this programme attempts to achieve international consensus on recommended health-based occupational exposure limits, in order to prevent impairment of the health of workers arising from toxic substances. This consensus will also have an impact on the harmonization of operational occupational exposure limits as decided upon by various national authorities. The Study Group recommends that WHO continue this programme by developing recommended health-based occupational exposure limits to toxic substances, selected according to the criteria mentioned in the report.

2. The Study Group recommends that the programme on health-based limits in occupational exposure should be carried out in close coordination with the WHO programme on the early detection of health impairment in workers exposed to occupational factors. In this way a highly important body of data on selected chemical substances will become available.

3. The Study Group concludes that the success of the WHO programme on health-based limits in occupational exposure depends on a multidisciplinary approach and on full participation of professionals from various disciplines—occupational health physicians and nurses, industrial hygienists, industrial toxicologists, and epidemiologists. The Group recommends that WHO continue to consult with these experts and with occupational health institutes all over the

world, in order to achieve a higher degree of effectiveness and application for the programme.

4. The Group recommends that WHO should initiate and activate epidemiological research in occupational health, with a view to increasing the potential of countries to carry out appropriate epidemiological studies on occupational exposure to harmful agents.